土 木 工 程 概 论

Introduction to Civil Engineering

黄 莺 主编

姚 尧 主审

中国建筑工业出版社

图书在版编目(CIP)数据

土木工程概论 = Introduction to Civil Engineering:
英文 / 黄莺主编. —北京: 中国建筑工业出版社, 2020.9 (2024.6重印)
ISBN 978-7-112-25439-2

Ⅰ.①土…　Ⅱ.①黄…　Ⅲ.①土木工程—高等学校—
教材—英文　Ⅳ.①TU

中国版本图书馆CIP数据核字(2020)第174706号

随着国民经济的迅速发展, 土木工程作为我国的一项基础建设工程也有了较大的飞跃。同时, 土木工程专业的求学人员和从业人员日益增多。土木工程概论能够帮助土木工程专业初学者了解本学科的基础知识, 对土木工程及其主要内容、基本理论、研究方法等有更深刻的认识。

Introduction to Civil Engineering (土木工程概论) 一书采用英文作为主要写作语言, 用英文系统地介绍土木工程的内容, 涉及土木工程基本概念、工程材料、建筑工程、土木工程、工程施工与管理、防灾减灾、土木工程信息化技术等内容。本书可帮助土木工程专业学生在学习专业课之前了解自己所学专业并对将来从事工作有一定认识, 也可帮助其他专业学生了解土木工程, 拓宽知识领域。本书采用英文为主要语言, 不但能够满足双语、全英文教学方式的需求, 帮助学生了解本专业的专业术语, 还可以作为土木类专业学生的英文阅读材料, 为学生日后科研学习及出国深造奠定基础, 提高学生专业知识能力。本书可作为土木类高校各类专业, 如土木工程、工程管理、交通工程、环境工程、机电工程、安全工程等, 本科生、专科生及研究生的双语教材或教学参考用书, 也可供土木建筑领域专业技术和管理人员学习参考。

责任编辑: 刘瑞霞
责任校对: 赵　菲

土木工程概论
Introduction to Civil Engineering
黄　莺　主编
姚　尧　主审
＊
中国建筑工业出版社出版、发行(北京海淀三里河路9号)
各地新华书店、建筑书店经销
北京科地亚盟图文设计有限公司制版
建工社（河北）印刷有限公司印刷
＊
开本: 　787×1092毫米　1/16　印张: 16¾　字数: 359千字
2020年11月第一版　　2024年6月第三次印刷
定价: **48.00**元
ISBN 978-7-112-25439-2
　　　　(36325)

序 言

　　土木工程是建造各类工程设施的科学技术的统称，它既指所进行的各类工程技术活动，也指工程建设的各种对象及工程设施。从我国古代的都江堰、大运河、紫禁城到如今的东方明珠、鸟巢、水立方等，这些世界著名的建筑体现着土木工程建造技术的不断革新。土木工程已逐渐成为了我国国民经济的支柱产业，对我国的经济发展起到了促进作用。随着科学技术的进步和工程实践的发展，土木工程这个学科也已发展成为内涵广泛、门类众多、结构复杂的综合体系。

　　Introduction to Civil Engineering（土木工程概论）一书，从科学概论的视角了解土木工程的综合性、社会性、实践性及其在技术、经济的统一性，并初步构建专业基础，树立专业思想。"土木工程概论"在整个土木工程专业知识体系中占有重要地位，教材内容丰富、涉及广泛，与土木工程专业内容的学习相互交织，并与土木工程实践联系紧密，知识更新紧跟工程发展；同时还兼具专业导论的功能，引导土木类专业初学者全面了解专业知识体系以及专业学习规律，激发专业荣誉感和责任感，为未来的专业学习奠定基础。本书内容既涉及工程材料、工程力学、工程图学等理论知识，又涵盖基础工程、建筑工程、道路与交通工程、桥梁工程、隧道与地下工程、水利工程、给排水工程、项目管理、防灾减灾等土木工程领域众多学科分支和研究方向，全面展示土木学科丰富的专业内涵，又合理引导大家认识土木工程专业，增强学习兴趣，帮助其初步建立专业思维。

　　Introduction to Civil Engineering（土木工程概论）一书作为一本双语课程教材，采用英文作为主要写作语言，用英文系统地介绍土木工程的内容，涉及土木工程基本概念、工程材料、建筑工程、土木工程、工程施工与管理、防灾减灾、土木工程信息化技术等内容。

　　本书由西安建筑科技大学黄莺主编，西安建筑科技大学赵平、西北工业大学陈昌宏、长安大学袁春燕副主编，西安建筑科技大学姚尧主审，刘梦茹统稿。研究生何易飞、孔繁盛、刘梦茹、魏晋果、熊文文、李嘉晨、宋睿参与了部分章节的编写。

　　由于编者水平有限，不足之处在所难免，敬请广大专家学者、同行和读者批评指正。在编写过程中，参考了一些同类教材及国内外研究学者的学术论文和学术专著，均在参考文献中列出，在此特向参考文献中提及的国内外各位学者表示衷心的感谢。

编　者

2020 年 7 月

Catalogue

CHAPTER 1   CIVIL ENGINEERING &CIVIL ENGINEERS   / 1

   1.1   What is Civil Engineering?   / 1

   1.2   Sub-disciplines of Civil Engineering   / 2

CHAPTER 2   CIVIL ENGINEERING MATERIALS   / 7

   2.1   The classification and development of CE materials   / 7

   2.2   Relationship between CE materials and engineering structures   / 13

   2.3   Main properties and characteristics of CE materials   / 15

   2.4   Natural stone and soil   / 16

   2.5   Timber   / 20

   2.6   Concrete and Reinforced Concrete   / 22

   2.7   Steel   / 26

   2.8   Asphalt   / 31

   2.9   Emerging CE Materials   / 32

   Questions   / 36

   Reference List   / 36

CHAPTER 3   CONSTRUCTION ENGINEERING   / 37

   3.1   Basic Components   / 37

   3.2   Truss Structure and Frame Structure   / 46

   3.3   Single and Multi-story Buildings   / 52

   3.4   High-rise and Ultra High-rise Buildings   / 60

   3.5   Special Structure   / 66

   3.6   Structure Calculation   / 71

   Questions   / 76

   Reference List   / 76

CHAPTER 4   CIVIL ENGINEERING STRUCTURES / 77

4.1   Road Engineering / 77

4.2   Urban Rail and Underground Railway / 82

4.3   Railway Engineering / 85

4.4   Airport Engineering / 88

4.5   Tunnel Engineering / 93

4.6   Bridge Engineering / 104

4.7   Development and Utilization of Subsurface / 115

4.8   Water Resources and Hydropower Engineering / 120

Question / 130

Reference List / 131

CHAPTER 5   CONSTRUCTION OF CIVIL ENGINEERING / 132

5.1   Earthwork Construction / 132

5.2   Foundation Construction / 135

5.3   Masonry Construction / 138

5.4   Reinforced Concrete Construction / 142

5.5   Prestressed Concrete Construction / 149

5.6   Installation Project / 153

5.7   Decoration Projects / 160

5.8   Steel Structure Construction / 164

5.9   Development Prospect of Construction Technology / 170

Questions / 172

Reference List / 173

CHAPTER 6   CONSTRUCTION MANAGEMENT / 174

6.1   Construction Procedures and Construction Codes / 174

6.2   Project Management / 185

6.3   Construction Organization / 201

Questions / 211

Reference List / 212

CHAPTER 7   HAZARD MITIGATION OF CIVIL ENGINEERING   / 213

    7.1   Introduction of Hazard   / 213

    7.2   Various Engineering Disasters and Prevention and Control Measures   / 215

    7.3   Engineering Structure Detection, Identification and Reinforcement   / 228

    7.4   The New Achievements and Development Trend of Disaster Prevention

        and Reduction   / 232

    Questions   / 234

    Reference List   / 235

CHAPTER 8   APPLICATION OF INFORMATION TECHNOLOGY IN

                CIVIL ENGINEERING   / 236

    8.1   Computer-aided Calculation   / 236

    8.2   Building Information Modeling (BIM)   / 244

    8.3   Information Construction   / 250

    8.4   Future of Civil Engineering IT   / 252

    Questions   / 256

    Reference List   / 257

CHAPTER 1
CIVIL ENGINEERING &CIVIL ENGINEERS

Civil engineering is a professional engineering discipline that deals with the design, construction, and maintenance of the physical and naturally built environment, including works like roads, bridges, canals, dams, and buildings. Civil engineering is traditionally broken into a number of sub-disciplines. It is the second-oldest engineering discipline after military engineering, and it is defined to distinguish non-military engineering from military engineering. Civil engineering takes place in the public sector from municipal through to national governments, and in the private sector from individual homeowners through to international companies. This chapter elaborates what Civil Engineering is.

1.1   What is Civil Engineering?

Civil engineering, the oldest of the engineering specialties, is the planning, design, construction, and management of the built environment. This environment includes all structures built according to scientific principles, from irrigation and drainage systems to rocket-launching facilities.

Civil engineers build roads, bridges, tunnels, dams, harbors, power plants, water and sewage systems, hospitals, schools, mass transit, and other public facilities essential to modern society and large population concentrations. They also build privately owned facilities such as airports, railroads, pipelines, skyscrapers, and other large structures designed for industrial, commercial, or residential use. In addition, Civil engineers plan, design, and build complete cities and towns, and more recently have been planning and designing space platforms to house self-contained communities.

The word "civil" derives from the Latin for citizen. In 1782, Englishman John Smeaton (FIG.1-1) used the term to differentiate his nonmilitary engineering work from that of the military engineers who predominated at the time. John Smeaton was also the first person who called himself as "civil engineer". He is often regarded as the "father of civil engineering." He had done a lot of work in the fields of civil engineering and mechanical engineering, and was responsible for the design of bridges, canals, ports and lighthouses. The Eddy stone Lighthouse (FIG.1-2) built by Smithton was also very famous. Since then, the term civil engineering has often been used to refer to engineers who build public facilities, although the field is much broader.

FIG.1-1   John Smeaton   **FIG.1-2**   Eddy stone Lighthouse

1.2   Sub-disciplines of Civil Engineering

Because it is so broad, civil engineering is subdivided into a number of technical specialties. Depending on the type of project, the skills of many kings of civil engineer specialists may be needed.

When a project begins, the site is surveyed and mapped by civil engineers who locate utility placement water, sewer, and power lines. Geotechnical specialists perform soil experiments to determine if the earth can bear the weight of the project. Environmental specialists study the project's impact on the local area: the potential for air and groundwater pollution, the project's impact on local animal and plant life, and how the project can be designed to meet government requirements aimed at protecting the environment. Transportation specialists determine what kind of facilities are needed to ease the burden on local roads and other transportation networks that will result from the completed project. Meanwhile, structural specialists use preliminary date to make detailed designs, plans, and specifications for the project. Supervising and coordinating the work of these civil engineer specialists, from beginning to end of the project, are the construction management specialists. Based on information supplied by the other specialists, construction management civil engineers estimate quantities and costs of materials and labor, schedule all work, order materials and equipment for the job, hire contractors and subcontractors, and perform other supervisory work to ensure the project is completed on time and as specified.

Throughout any given project, Civil engineers make extensive use of computers. Computers are used to design the project's various elements (computer-aided design, or CAD) and to manage it. Computers are a necessity for the modern civil engineers because they permit the engineer to efficiently handle the large quantities of data needed in determining the best way to construct a project.

Structural Engineering

In this specialty, civil engineers plan and design structures of all types, including bridges, dams, power plants, supports for equipment, special structures for offshore projects, the United States space program, transmission towers, giant astronomical and radio telescopes, and many other kinds of projects. Using computers, structural engineers determine the forces a structure must resist: its own weight, wind and hurricane forces, temperature changes that expand or contract construction materials, and earthquakes. They also determine the combination of appropriate materials, steel, concrete, plastic, stone, asphalt, brick, aluminum, or other construction materials.

Water Resources Engineering

Civil engineers in this specialty deal with all aspects of the physical control of water. Their projects help prevent floods, supply water for cities and for irrigation, manage and control rivers and water runoff, and maintain beaches and other waterfront facilities. In addition, they design and maintain harbors, canals, and locks, build huge hydroelectric dams and smaller dams and water impoundments of all kinds, help design offshore structures, and determine the location of structures affecting navigation.

Geotechnical Engineering

Civil engineers who specialize in this field analyze the properties of soils and rocks that support structures and affect structural behavior. They evaluate and work to minimize the potential settlement of buildings and other structures that stems from the pressure of their weight on the earth. These engineers also evaluate and determine how to strengthen the stability of slopes and fills and how to protect structures against earthquakes and the effects of groundwater.

Environmental Engineering

In this branch of engineering, civil engineers design, build, and supervise systems to pro-

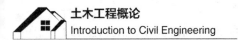

vide safe drinking water and to prevent and control pollution of water supplies, both on the surface and underground. They also design, build, and supervise projects to control or eliminate pollution of the land and air. These engineers build water and waste water treatment plants, and design air scrubbers and other devices to minimize or eliminate air pollution caused by industrial processes, incineration, or other smoke producing activities. They also work to control toxic and hazardous wastes through the construction of special dump sites or the neutralizing of toxic and hazardous substances. In addition, the engineers design and manage sanitary landfills to prevent pollution of surrounding land.

Transportation Engineering

Civil engineers working in this specialty build facilities to ensure safe and efficient movement of both people and goods. They specialize in designing and maintaining all types of transportation facilities, highways and streets, mass transit systems, railroads and airfields, ports and harbors. Transportation engineers apply technological knowledge as well as consideration of the economic, political, and social factors in designing each project. They work closely with urban planners, since the quality of the community is directly related to the quality of the transportation system.

Pipeline Engineering

In this branch of civil engineering, engineers build pipelines and related facilities which transport liquids, gases, or solids ranging from coal slurries (mixed coal and water) and semi-liquid wastes, to water, oil, and various types of highly combustible and noncombustible gases. The engineers determine pipeline design, the economic and environmental impact of a project on regions it must traverse, the type of materials to be used-steel, concrete, plastic, or combinations of various materials-installation techniques, methods for testing pipeline strength, and controls for maintaining proper pressure and rate of flow of materials being transported. When hazardous materials are being carried, safety is a major consideration as well.

Construction Engineering

Civil engineers in this field oversee the construction of a project from beginning to end. Sometimes called project engineers, they apply both technical and managerial skills, including knowledge of construction methods, planning, organizing, financing, and operating construction projects. They coordinate the activities of virtually everyone engaged in the work: the surveyors, workers who lay out and construct the temporary roads and ramps, excavate for the foundation, build the forms and pour the concrete, and workers who build the steel framework. These engineers also make regular progress reports to the owners of the structure.

Community and Urban Planning

Those engaged in this area of civil engineering may plan and develop communities within a city or entire cities. Such planning involves far more than engineering consideration; environmental, social, and economic factors in the use and development of land and natural resources are also key elements. These civil engineers coordinate planning of public works along with private development. They evaluate the kinds of facilities needed, including streets and highways, public transportation systems, airports, port facilities, water-supply and wastewater-disposal systems, public buildings, parks, and recreational and other facilities to ensure social and economic as well as environmental well-being.

Photogrammetry, Surveying, and Mapping

The civil engineers in this specialty precisely measure the Earth's surface to obtain reliable information for locating and designing engineering projects. This practice often involves high-technology methods such as satellite and aerial surveying and computer-processing of photographic imagery. Radio signals from satellites, scans by laser and sonic beams, are converted to maps to provide far more accurate measurements for boring tunnels, building highways and dams, plotting flood control and irrigation projects, locating subsurface geologic formations that may affect a construction project, and a host of other building uses.

Other specialties. Two additional civil engineering specialties that are not entirely within the scope of civil engineering but are essential to the discipline are engineering management and engineering teaching.

Engineering Management

Many civil engineers choose careers that eventually lead to management. Others are able to start their careers in management positions. The civil engineer-manager combines technical knowledge with an ability to organize and coordinate worker power, materials, machinery, and money. These engineers may work in government-municipal, county, or province.

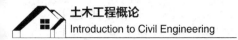

Engineering Teaching

The civil engineer who chooses a teaching career usually teaches both graduate and undergraduate students in technical specialties. Many teaching civil engineers engage in basic research that eventually leads to technical innovations in construction materials and methods. Many also serve as consultants on engineering projects, or on technical boards and commissions associated with major projects.

CHAPTER 2
CIVIL ENGINEERING MATERIALS

Civil engineering materials refer to the general term for various materials used in the project of civil engineering, which are the material basis for all construction projects. The civil engineering materials industry promotes the development of the construction industry and is one of the important basic industries of the national economy.

2.1 The classification and development of CE materials

2.1.1 The classification of civil engineering materials

At present, the commonly used engineering materials are: stone, wood, brick, tile, cementing material, mortar, asphalt, asphalt mixture, steel, reinforced concrete, and some auxiliary materials and new materials. There are roughly two basic classification methods, one is to classify by material composition, and the other is to classify by material's role.

When classified by material composition, engineering materials can be divided into two categories: non-metallic materials and metallic materials, and non-metallic materials can be divided into inorganic materials and organic materials. The main components of materials such as sand, stone, brick, tile, lime, cement, mortar, concrete, and inorganic fibers (such as glass fiber) are inorganic materials. It should be noted that the concrete here refers to concrete in a narrow sense, that is, cement, sand and water are mixed in a certain proportion. Of course, it is not excluded that inorganic materials contain a certain amount of organic impurities or artificially added a small amount of organic substances to improve the performance of certain materials, but in general, their main components are still inorganic.

In the category of organic materials, wood, asphalt, paint, organic fibers, and plastic are common (FIG.2-1). These materials can be used as structural materials, such as wood used in wooden structures, and organic fibers added to fiber concrete. However, currently organic materials are

(a) Sand   (b) Stone

(c) Brick   (d) Tile

FIG.2-1   The inorganic materials

mostly used for non-structural parts, such as paints, chemical fiber products and plastic pipes are commonly used in decoration (FIG.2-2).

As for metal materials, they can be divided into two categories: ferrous metals and non-ferrous metals. Ferrous metals mainly refer to iron and its alloys. Pig iron, cast iron, carbon steel, and alloy steel are all ferrous metals. They are the most used metals in the world at present, accounting for almost 95% of the total metal usage. Of course, this is also the most used metal in our engineering, steel bars and section steel in the structure and the materials used for the scaffolding and steel formwork during construction are also ferrous metals (FIG.2-3). And non-ferrous metals such as aluminum, lead, copper, and zinc also have a certain amount of demand in the engineering project, but the overall usage is relatively small and their economic costs are relatively higher.

When classified according to the role of materials, the engineering materials are usually divided into load-bearing materials (also called structural materials), enclosure materials, decorative materials, and gelling materials. Load-bearing materials refer to materials that mainly bear various external effects suffered by the structure, so these materials are mainly used in structural parts such as wall, columns, beams, slabs and foundations. Materials such as steel, concrete, masonry, and wood are Load-bearing materials commonly used in construction currently or in the

(a) Wood   (b) Asphalt

(c) Paint   (d) Plastic

FIG.2-2   The organic materials

FIG.2-3   The scaffolding and steel formwork

past. Enclosure materials refer to such materials that can maintain the function of separating spaces and passages, including bricks, blocks, lightweight concrete, and other wall materials. These materials can play a role in dividing and connecting spaces. Decorative materials refer to materials that can create a beautiful and comfortable environment, the stone, wood, glass and paint are belong to this material. Cementing materials refer to a class of materials that have a certain cementation effect and can bind loose particles into a whole, such as cement, lime, gypsum, asphalt and so on. The materials with different functions are shown in FIG.2-4.

(a) Section steel   (b) Blocks

(c) Glass   (d) Gypsum

FIG.2-4   The materials with different functions

2.1.2   The development of civil engineering materials

Early civil engineering materials were mainly stone, brick and wood. Stone is one of the oldest civil engineering materials. World-famous sites such as the Pyramid of Khufu and Parthenon (FIG.2-5) are all built of stone. Stone is mainly used for masonry under pressure, this is due to the high compressive strength of stone, which is far superior to its other strength indicators.

FIG.2-5   Brick structure——The Pyramid of Khufu and the Parthenon

Brick, as a material of masonry, has a long history of use in China. The Chinese ancient buildings are known as Qin bricks and Han tiles. In the past, bricks were mainly used as load-bearing structural materials for masonry of walls, columns and foundations (FIG.2-6). According to different production processes, bricks can be divided into sintered bricks and non-sintered bricks. Sintered bricks have been widely used. But at present, sintered clay bricks have been ordered to ban because of the large use of them that destroyed fields.

Wood is also an ancient engineering material. The wood in construction works can be used for thousands of years if properly treated. Wood has many advantages, such as light weight and high strength; easy processing; high elasticity and toughness; ability to withstand shock and vibration. However, there are also some disadvantages of wood, such as flammability, easy to corrosion and large structural deformation. However, these disadvantages are greatly improved after processing. In ancient times, wood occupies a very important position in engineering construction. The Bell Tower which is the landmark of Xi'an is a brick-wood structure (FIG.2-7), it has a history of five or six hundred years.

FIG.2-6 Brick structure——The Great Wall   **FIG.2-7** Brick-wood structure——The Bell Tower of Xi'an

The ancient Chinese techniques of building wooden structures can be described as peaking, such as complex and clever dougong, which have the function of load bearing and decoration, as well as the ingenious connection technology of wooden structure nodes——mortise and tenon joint, which reflects the wisdom of our ancestors (FIG.2-8).

With the advent of the industrial age, cement, steel and concrete have rapidly developed as a new generation of building materials. Cement (FIG.2-9) is a powdery hydraulic gelling material, when mixed with water, it can be coagulated and hardened in air and water. This material can cement other materials into a whole to form a hard material. Although it has been less than 200 years since its invention, cement is already the material which has the largest output and quantity in civil engineering today. Cement and its cement-based materials such as mortar and concrete are commonly used in various civil engineering, especially reinforced concrete structures.

FIG.2-8   The dougong and the mortise and tenon joint

Concrete (FIG.2-10) is currently the most popular engineering material. Broadly speaking, concrete refers to solid materials formed by mixing various organic, inorganic, natural, and artificial gelling materials with granular or fibrous fillers. In a narrow sense, concrete refers to artificial stone made from cementitious material, aggregate and water in a certain ratio, formed by stirring and vibration, and cured under certain conditions. In general, artificial stone made from cement, coarse aggregate (that is, crushed stone or pebble), fine aggregate (sand), and water is ordinary concrete.

FIG.2-9   Cement

FIG.2-10   Concrete

Steel (FIG.2-11) has been used to build various houses and bridges since the early 19th century. Compared with concrete materials, steel has high strength, superior mechanical properties, convenient construction, and can be recycled. However, at the same time, because the steel is poor in corrosion resistance, high temperature resistance, fire resistance, and the high cost of steel structure construction, so steel structures are mainly used in ultra-high or long-span structures at present. The steel for civil engineering is mainly various section steel, steel bars, steel wires, and steel strands.

With the development of materials technology, a large number of new materials have appeared, such as some polymer materials or composite materials. Some fiber materials, such as carbon fiber, have the characteristics of high temperature resistance, corrosion resistance and

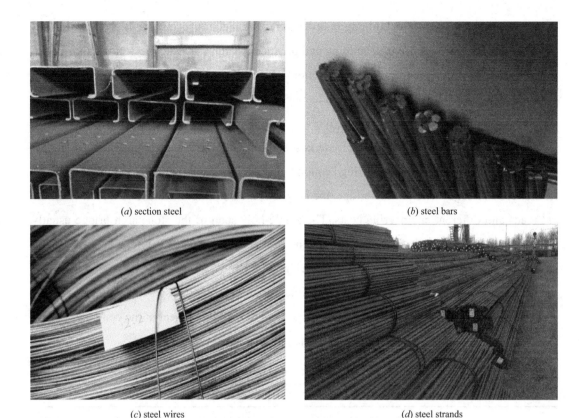

(*a*) section steel   (*b*) steel bars

(*c*) steel wires   (*d*) steel strands

FIG.2-11   Various steel materials

high tensile strength. It can be added to concrete to form carbon fiber concrete or can be used alone. Carbon fiber cloth is often used to reinforce existing structural members and handle structural cracks. In addition, there are high-molecular composite films such as those used in membrane structures. For example, the only seven-star hotel in the world is the Sailing Hotel in Dubai (FIG.2-12), its outer sailing ship shape is a double-layer membrane structure.

FIG.2-12   The Sailing Hotel in Dubai

2.2   Relationship between CE materials and engineering structures

The requirements of civil engineering for civil engineering materials are that they must have sufficient strength to withstand the design load safely. Its surface density is light, to reduce the load on the lower part of the structure and foundation. The materials has a durability that is compatible with the life of the building in order to reduce maintenance costs. It used for decoration should be able to beautify the building and produce a certain artistic effect. It used in special parts should have corresponding special functions. At the same time, civil engineering materials should save resources and energy as much as possible in the production process to achieve sustainable development.

The selection of civil engineering materials is closely related to the engineering design proposal, construction scheme, engineering economy and performance. Different engineering materials directly affect the span of the bridge, the height of the building, the maximum load of different roads and different engineering costs. The single-hole span of the stone arch bridge usually does not exceed 20m, while the single span of the cable-stayed suspension bridge has already exceeded 1000m. The main span of the Akashi Bridge in Japan was 1991m, while the main span of the Gibraltar Strait Bridge is 5000m. The maximum load capacity of asphalt pavement in China shall not exceed 10 tons, while the vertical design load of the main wheel of the concrete runway may vary greatly depending on the aircraft model, generally exceeding 20 tons.

Table 2-1   Scope of Application of Building Materials

Masonry	Multi-storey houses, hotels, etc.
Concrete	Multi-story residential buildings, public buildings (office, business, research, education, hospitals, etc.)
Steel	Large structure, high-rise and super high-rise structure
Timber	Single or multi-storey residential and public buildings, etc.

In engineering construction, different materials require different processing methods and construction techniques. For example, the processing techniques of steel beams, concrete beams and wooden beams are completely different. There are large differences in the treatment methods for soft soil foundations, collapsible loess foundations and non-uniformly expansive soil foundations.

The material has a decisive impact on the project cost. Although block materials and concrete materials can be used in multi-storey houses, the project cost difference between the two houses below 7 floors is from 30% to 40%. From another point of view, the schemes of using steel and concrete are more expensive, and its seismic behavior is superior to the masonry structure. Therefore, the selection of engineering materials should be reasonably considered in safety, economy, adaptability, durability and reality of the project, and should be scientific according to local conditions.

The material determines the structural form. In underground engineering, we can clearly understand the impact of different soil materials on underground structures and foundation forms.

For example, in first and second class of site soil with a large bearing capacity, a masonry structure or a frame structure on a lower number of layers may apply a shallow foundation such as an independent foundation or a strip foundation; while on the four types of soft soil layers, pile foundation or full house foundation is only used in this condition.

The life of the project is constrained by materials. In roads and traffic, different materials can affect the service life of the pavement. (Table 2-2)

Table 2-2　Design Life of Pavement Structure

Road Grade	Bitumen	Cement Concrete	Concrete Block	Stone
Expressway	15	30	—	—
Main Road	15	30	—	—
Distributor Road	10	20	—	—
Branch Road	8	15	10	20

In the history of earthquake damage, it can be observed that different materials have a great impact on the seismic performance of civil engineering. In the Wenchuan earthquake in Sichuan and the Yushu earthquake in Qinghai, which occurred in recent years in China, eighty percent of buildings with earthquake damage are masonry structures, and the seismic damage rate of concrete structures is about 10%, while the seismic damage rate of steel structures is less than 3%.

The impact of engineering materials on civil engineering is closely related to the nature of the material itself. In the following, we discuss the properties of several typical building materials in civil engineering.

2.3　Main properties and characteristics of CE materials

Civil engineering materials have many properties, such as physical properties, chemical properties, mechanical properties and functional properties. Its physical and chemical properties are also referred to as the basic properties of the material.

The physical properties of the material mainly include density, elastic modulus, Poisson's ratio, porosity, water content, etc. The chemical properties of the material mainly include chemical components, hydrophilicity, water repellency, aging, corrosion and the like. The influence of the basic properties of materials on other properties of materials is an important research direction in materials science. The classification of materials usually depends on the basic properties of the material. For example, density divides stone into saturated soil and unsaturated soil; the carbon content is divided into low carbon steel, medium carbon steel and high carbon steel. The chemical composition of the steel determines whether there is a phenomenon of the hot or low temperature brittleness in structural steel.

The mechanical properties of a material refer to the strength, deformation and destructibility of the material under various loads. Material strength is the maximum load that a material is subjected to per unit area. The deformation capacity of a material is referred to the plasticity of the material in engineering. A material with poor deformability, without warning, failure to warn, and avoiding hazards caused by damage, is called brittle material. In the same kind of material, strength and plasticity seem to be incompatible. The newly developed high-strength concrete is stronger than common concrete, but the plasticity is worse; the same is true for steel. The anti-fatigue ability of a material indicates the ability of the material to function properly under repeated application of dynamic loads, which is usually measured by fatigue strength. The pavement material is repeatedly crushed by the vehicle, so the anti-fatigue ability of the material determines the normal service life of the road. The functional properties of civil engineering materials mainly include fire resistance, heat-shielding performance, corrosion resistance, etc. It is the basis for the selection of functional materials. For example, the choice of fire-resistant materials in buildings is primarily concerned with the thermal properties of the materials at high temperatures, and the primary concern of insulation materials is the thermal conductivity of the materials.

2.4 Natural stone and soil

2.4.1 Natural stone

1. Classification of stone and common stone

Rocks are processed by mechanical or manual methods, and by various block or loose materials obtained without processing, which are collectively referred to as natural stone. Natural

stone is one of the oldest building materials used by humans. There are many famous stone buildings in the world, such as the ancient Egyptian pyramids, the Colosseum in ancient Rome, the Leaning Tower of Pisa in Italy, the Zhaozhou Bridge in China and so on. Natural stone has a wide range of applications in water resources and hydropower, civil engineering, roads, bridges and other projects.(FIG.2-13)

FIG.2-13 The structure of Zhaozhou Bridge by natural stone

The characteristics of natural stone are mainly high compressive strength, good durability, wear resistance, beautiful appearance, and wide distribution for easy access to local materials. The disadvantages of natural stone are mainly brittleness, heavy weight, difficult mining and processing, and poor seismic performance of its structure. Rocks are divided into three major categories: magmatic rocks, sedimentary rocks and metamorphic rocks due to different geological conditions. They have different mineral compositions and different structural characteristics, and their architectural properties and scope of application are mainly determined by these characteristics.

2. Technical nature of stone

Due to the different parts and engineering properties, the technical requirements of the stone are also different. For example, stone materials applied to foundations, bridges, tunnels and masonry projects generally require high compressive strength, frost resistance and water resistance; stones used in architectural decoration, in addition to a certain degree of resistance to frost resistance and water resistance, have higher requirements on the density and wear resistance of stone.

FIG.2-14   The properties of natural stone

- **Surface density**

The surface density of rock is related to its mineral composition and porosity. The surface density of dense rock is 2500~3100kg/m³, such as granite; the surface density of rocks with a large porosity is 500~1700kg/m³, such as pumice. Rocks with a surface density greater than 1800 kg/m³ are heavy stones, and rocks with an surface density of less than 1800 kg/m³ are pumice. The stone processed by heavy stone can be used for the foundation of the structure, the ground, the decorative veneer, the wall, the bridge and the dam, etc. The stone processed by the pumice is mainly used as the wall material.

- **Strength**

The mechanical properties of masonry stone are mainly considering its compressive strength. The compressive strength of the cube test piece with a side length of 70 mm indicates the strength grade of the masonry stone, and the compressive strength is the average value of the fracture strength of the three test pieces. The strength grade of natural stone is classified into 7 classes: MU20, MU30, MU40, MU50, MU60, MU80 and MU100. The strength of dense rock is high, especially its compressive strength can be as high as 250~350MPa, generally 40~100MPa. The tensile strength of the rock is not high, and the tensile strength of the rock is usually reflect-

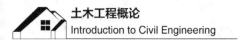

ed by the flexure strength. Generally, the tensile strength of dense rock is 1/50~1/4 of the compressive strength.

- **Waterproof**

According to the softening coefficient, the water resistance of rock is divided into high, medium and low. The rock with softening coefficient greater than 0.90 is high water resistant rock, and the rock with softening coefficient of 0.70~0.90 is medium water resistant rock. Stones with a softening coefficient below 0.60 are generally not approved for use in important buildings, such as in warm climates, or when the stone still has a high compressive strength after water saturation, it can be used with caution.

- **Frost Resistance**

Frost resistance refers to the ability of rocks to resist freeze-thaw damage, which is an important indicator of rock durability. The frost resistance of rocks is closely related to the water absorption rate. Generally, rocks with low water absorption have good frost resistance. In addition, the frost resistance of stone depends on its mineral composition, grain size and distribution uniformity, cementation of natural cement, porosity and other properties. After a predetermined number of freeze-thaw cycles of rock, if there is no through crack and the mass loss does not exceed 5% and the strength loss does not exceed 25%, the frost resistance of rocks under Water-saturated Condition is acceptable.

- **Weathering Resistance**

The cracking of rocks caused by water, ice and chemical factors is called weathering of rocks. The strength of rock weathering resistance is related to its mineral composition, structure and structural state. All the cracks on the rock make it easy for water to enter, causing it to gradually break down. Granite and the like have good weather resistance. The weatherproofing measures mainly include polishing the stone to prevent surface water, and coating the surface with silicone.

- **Process Performance**

The process performance of building stone represents the difficulty and possibility of stone mining and processing, including processability, polishing and drilling resistance. The processing performance is extremely important for the stone used for architectural decoration, which directly affects the decorative effect of the stone.

2.4.2 Soil

Every work of construction in civil engineering is built on soil or rock and in many instances these are also the raw materials of construction. So questions about the properties of soil are as important as those about other constructional materials. There questions include:

- Will a given soil provide permanent support for a proposed building?
- Will a given soil compress (or swell) due to application of load from a proposed building, and by what amounts and at what rates?
- What will be the margin of safety against failure or excessive or unequal compression of the soil?
- Are natural or constructed soil slopes stable or likely to slide?
- What force is exerted on a wall when a given soil is packed against it?
- At what rate and to what pattern will water flow through a soil?
- Can a given soil be improved in any way by treatment or admixture?

All these require an understanding of soils as materials and of the mechanics of materials. Absolute answers are unobtainable and the engineers have to practice the art as the science of geo-technology.

Anyway, to learn about the stabilization and strengthening measures of soil is very important for civil engineering students.

In the wider sense, soil stabilization and strengthening measures are terms used for improvement of soils either as they exist in-situ or when laid and densified as fill. The purpose of stabilization is to make a soil less pervious, less compressible or stronger, or all of these. This is often achieved by injecting a gelling or hardening fluid into the pore spaces in soil. In suitable soils stabilization may also be achieved by introducing admixtures and then applying mechanical work or vibration to densify the materials.

FIG.2-15 Soil reinforcement

Most of the construction in civil engineering is built on soil or rock, and the soil as the foundation of the building plays a vitally important role in ensuring the durability of the building. Otherwise, the soil as the construction material is widely applied in architectures. From the beacon towers, tombs and the ruins of the old city that were preserved in ancient times, it can be seen that the ancients used raw soil to build buildings. Earth architecture are widely distributed throughout the world. In 1981, the Centre National d'Art et de Culture, Georges Pompidou hosted exhibitions of earth-going architecture around the world. About one-third of the world's population lived in earth architecture in the early 21st century. Due to geographical conditions,

lifestyles, historical traditions, and ethnic customs, the earth construction in each region has its own characteristics in construction technology and architectural style, which has become an integral part of the architectural culture of various countries.

As a building material, soil can be taken locally, easy to construct, low in cost, cool in summer and warm in winter, saving energy, and it is also natural, which is conducive to environmental protection and ecological balance. Therefore, this ancient type of building still has vitality. However, all kinds of earth buildings have shortcomings, limited layout, insufficient sunshine, poor ventilation and humidity, and need to be improved.

2.5 Timber

Timber is an ancient engineering material. For example, the Liuhe Tower, located on the banks of the Qiantang River in Hangzhou, uses timber as the structures and has a history of thousands of years. The mechanical properties of timber are extremely unique, and the strength of the timber (the direction of the force parallel to the direction of the fiber) is much higher than the strength of the transverse stripes (the direction of the force perpendicular to the direction of the fiber). Therefore, timber is well suited to withstand tensile and bending moments. The timber is light, high-strength, easy to process (such as sawing, planning, drilling, etc.), has good elasticity and toughness, can withstand shock and vibration, and has excellent seismic performance, and is widely used in Japan. As a building material, timber is also the only natural resource that can be recycled in the four major building materials "steel, cement, plastic and timber". It is also a green and ecological building material. The timber has low conductivity and thermal conductivity and can be used as insulation material; the timber grain is beautiful which is fit for decoration.

FIG.2-16 "Dwelling" House with Wood (built by XAUAT)
(Resource: College of Architecture, XAUAT)

With the rapid development of various constructions, the demand for timber has become large, while the growth of trees is relatively slow, resulting in great contradiction between supply and demand. The research team—Bhavna Sharma, Ana Gatóo, Maximilian Bock, Michael Ramage, Department of Architecture, University of Cambridge, Cambridge, UK— found the ideal

FIG.2-17   Glue-laminated bamboo products

material for bamboo, which can be recognized as a substitute for timber. Their research indicates that bamboo is a fast-renewable material that has many applications in architecture. Engineered bamboo products result from processing the raw bamboo culm into a laminated composite, similar to glue-laminated timber products. These products allow the material to be used in standardized sections and have less inherent variability than the natural material. The present work investigates the mechanical properties of two types of commercially available products— bamboo scriber and laminated bamboo sheets—and compares these to timber and engineered timber products. It is shown that engineered bamboo products have properties that are comparable to or surpass that of timber and timber-based products. Potential limitations to use in structural design are also discussed. The study contributes to a growing body of research on engineered bamboo and presents areas in which further investigation is needed.

However, timber and bamboo have disadvantages, such as uneven structure, anisotropy, moisture absorption and water absorption, resulting in large swelling, shrinkage deformation, flammability, perishability, and the like.

FIG.2-18   The rotten wood

2.6　Concrete and Reinforced Concrete

2.6.1　Introduction to Concrete and Reinforced Concrete

Concrete is made of cemented material, aggregate and water in a certain proportion, and is formed by mixing and vibrating to form an artificial stone which is cured under certain conditions. There are many types of concrete: according to different cement materials, it is divided into ordinary concrete, asphalt concrete, gypsum concrete and polymer concrete; according to the density, it is divided into heavy concrete, ordinary concrete, light concrete; according to different functions, it is divided into structural concrete, road concrete, hydraulic concrete, heat-resistant concrete, acid-resistant concrete and radiation-proof concrete; according to different construction techniques, it is divided into shotcrete, pumped concrete, vibrating grouting concrete, etc.

Since time immemorial, Chinese, Egyptians and Romans have used concrete, calcined clay, calcined gypsum and lime mountain ash as cementitious materials. However, concrete made of gas-hardening cement materials such as lime and gypsum has the disadvantage of being incapable of water. Although lime and volcanic ash have certain hydraulic properties, their mechanical properties and durability are far from meeting the requirements of human civil engineering materials. After 1824, concrete with cement as a cementitious material began to appear, followed by reinforced concrete and prestressed concrete in 1850 and 1928. Concrete has been widely used since then. At present, it is the world's largest and most widely used civil engineering material.

Concrete has the advantages of high compressive strength, good durability and a wide range of strength grades. At the same time, it is rich in raw materials, simple in production process and low in price, so its use is increasing. However, concrete has the disadvantages of large self-weight, low tensile strength, and easy cracking.

At the 1867 Paris World Expo, a French gardener, Joseph Monier, presented his invention——a reinforced concrete pot and a railway sleeper. After that, he continued to search for new materials and obtained patents including reinforced concrete pots and reinforced concrete beams that were applied to highway guardrails. He made it possible to find that reinforcing steel is placed in the tensile zone and the easily crackable area of the concrete member, which can greatly improve the tensile strength and bending strength, and can limit the occurrence and development of cracks. His invention was quickly applied to the construction sector. In 1875, the world's first reinforced concrete structure, designed by Monier, was built at the Castle of Chazelet, Paris, which is a new era in the history of human architecture.

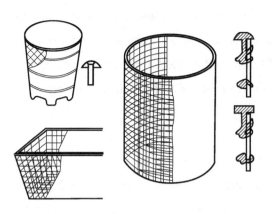

Reinforced Concrete Pots innovated by Monier

The world's first RC structure designed by Monier

FIG.2-19 RC by Monier

Reinforced concrete structures have been used extensively in the engineering community since 1900. In 1928, a new type of reinforced concrete structure, prestressed reinforced concrete, appeared and was widely used in engineering practice after the Second World War. The invention of reinforced concrete and the application of steel in the construction industry in the mid-19th century made it possible to construct tall buildings and long-span bridges.

FIG.2-20 The prestressed reinforced concrete

2.6.2 Performance of concrete

1. Fire safety

Concrete buildings are more resistant to fire than those constructed using steel frames, since concrete has lower heat conductivity than steel and can thus last longer under the same fire conditions. Concrete is sometimes used as a fire protection for steel frames, for the same effect as above. As a fire shield, for example, concrete can also be used in extreme environments like a missile launch pad.

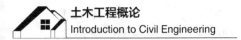

Options for non-combustible construction include floors, ceilings and roofs made of cast-in-place and hollow-core precast concrete. For walls, concrete masonry technology and Insulating Concrete Forms (ICFs) are additional options. ICFs are hollow blocks or panels made of fire-proof insulating foam that are stacked to form the shape of the walls of a building and then filled with reinforced concrete to create the structure.

2. Earthquake safety

As discussed above, concrete is very strong in compression, but weak in tension. Larger earthquakes can generate very large shear loads on structures. These shear loads subject the structure to both tensile and compressional loads. Concrete structures without reinforcement, like other unreinforced masonry structures, can fail during severe earthquake shaking. Unreinforced masonry structures constitute one of the largest earthquake risks globally. These risks can be reduced through seismic retrofitting of at-risk buildings, (e. g. school buildings in Istanbul, Turkey).

2.6.3 Application of concrete and reinforced concrete in construction

Concrete is one of the most durable building materials. It provides superior fire resistance compared with wooden construction and gains strength over time. Structures made of concrete can have a long service life. Concrete is used more than any other artificial material in the world. As of 2006, about 7.5 billion cubic meters of concrete are made each year, more than one cubic meter for every person on Earth.

1. Mass structures

Due to cement's exothermic chemical reaction while setting up, large concrete structures such as dams, navigation locks, large mat foundations, and large breakwaters generate excessive heat during hydration and associated expansion. To mitigate these effects post-cooling is commonly applied during construction. An early example at Hoover Dam used a network of pipes between vertical concrete placements to circulate cooling water during the curing process to avoid damaging overheating. Similar systems are still used; depending on volume of the pour, the concrete mix used, and ambient air temperature, the cooling process may last for many months after the concrete is placed. Various methods are also used to pre-cool the concrete mix in mass concrete structures.

Another approach to mass concrete structures that minimizes cement's thermal byproduct is the use of roller-compacted concrete, which uses a dry mix that has a much lower cooling requirement than conventional wet placement. It is deposited in thick layers as a semi-dry material then roller compacted into a dense, strong mass.

FIG.2-21　The Hoover Dam
(https://www.uniqueway.com)

2. Surface finishes

Raw concrete surfaces tend to be porous, and have a relatively uninteresting appearance. Many different finishes can be applied to improve the appearance and preserve the surface against staining, water penetration and freezing.

Examples of improved appearance include stamped concrete where the wet concrete has a pattern impressed on the surface, to give a paved, cobbled or brick-like effect, and may be accompanied with coloration. Another popular effect for flooring and table tops is polished concrete where the concrete is polished optically flat with diamond abrasives and sealed with polymers or other sealants.

Other finishes can be achieved with chiselling, or more conventional techniques such as painting or covering it with other materials.

The proper treatment of the surface of concrete, and its characteristics, is an important stage in the construction and renovation of architectural structures.

FIG.2-22　Colored Concrete and Polished Concrete

3. Prestressed structures

Prestressed concrete is a form of reinforced concrete that builds in compressive stresses during construction to oppose tensile stresses experienced in use. This can greatly reduce the weight of beams or slabs, by better distributing the stresses in the structure to make optimal use of the reinforcement. For example, a horizontal beam tends to sag. Prestressed reinforcement along the bottom of the beam counteracts this. In pre-tensioned concrete, the prestressing is achieved by using steel or polymer tendons or bars that are subjected to a tensile force prior to casting, or for post-tensioned concrete, after casting.

More than 55000 miles (89000 km) of highways in the United States are paved with this material. Reinforced concrete, prestressed concrete and precast concrete are the most widely used types of concrete functional extensions in modern days.

4. Roads

Concrete roads are more fuel efficient to drive on[6], more reflective and last significantly longer than other paving surfaces, yet have a much smaller market share than other paving solutions. Modern-paving methods and design practices have changed the economics of concrete paving, so that a well-designed and placed concrete pavement will be less expensive on initial costs and significantly less expensive over the life cycle. Another major benefit is that pervious concrete can be used, which eliminates the need to place storm drains near the road, and reducing

the need for slightly sloped roadway to help rainwater to run off. No longer requiring discarding rainwater through use of drains also means that less electricity is needed (more pumping is otherwise needed in the water-distribution system), and no rainwater gets polluted as it no longer mixes with polluted water. Rather, it is immediately absorbed by the ground.

FIG.2-23 The colored concrete road

2.7   Steel

2.7.1   Performance of Steel

From the beginning of the 19th century, mankind began to use steel for the construction of bridges and houses. By the middle of the 19th century, the variety, specifications and production scale of steel products have increased substantially, the strength has been continuously improved, and the processing technologies such as cutting and joining of steel have been greatly developed. It is widely used in various structural projects such as railways, bridges, and construction projects, and plays an important role in national economic construction. Steel for civil engineering refers to various profiles (such as round steel, angle steel, I-beam, etc.) used for steel structures, steel plates, pipes, and various steel bars and steel wires used in reinforced concrete.

(*a*) Steel Railway

(*b*) Steel Bridge

(*c*) Steel House

(*d*) Steel Workshop

FIG.2-24 The steel structure

The quality of steel is uniform and compact. The tensile strength, compression resistance, bending resistance and shear strength are very high. It can withstand large impact and vibration loads at normal temperature, and has certain plasticity and very good toughness. Steel has good processing properties and can be cast, forged, welded, riveted and cut for easy assembly. In addition, the properties of the steel can be changed or controlled over a wide range by a heat treatment method.

Compared with materials such as stone and concrete, the steel has high tensile strength and compressive strength, and has good deformation ability and easy processing performance. The seismic performance of steel structures is extremely superior. Due to its high strength, the cross-section dimensions of steel structural members can be longer, higher and slimmer than other materials, greatly expanding the span and height of modern engineering, making modern engineering a light and flexible visual effect.

The disadvantage of steel is that it is easy to rust, so it is necessary to do rust and corrosion protection of steel structure. The steel will soften in the fire. The collapsed World Trade Center in the "9/11 terrorist attacks in the United States" was caused by the fire of the aircraft striking the building, causing the steel to soften and the structure collapse. Therefore, fire protection of steel structures is equally important.

FIG.2-25　Steel component processing　　　**FIG.2-26**　World Trade Center in terrorist attacks

2.7.2　Fire Protection of Steel

In the general building structure, the steel works under normal temperature conditions, but for structures that may be exposed to high temperature for a long period of time, or when encountering special conditions such as fire, the influence of temperature on the properties of the steel must be considered. Moreover, the effect of high temperature on performance cannot be assessed simply by stress-strain relationship, while it must be combined with temperature and high temperature duration. Generally, the creep phenomenon of steel is more pronounced with increasing temperature, and creep causes stress relaxation. In addition, since the grain boundary strength is lower than the grain strength at high temperature, the sliding of the grain boundary plays an important role in the influence of the microcrack, and the crack continuously expands under the action of the tensile stress to cause fracture. Therefore, as the temperature increases, its permanent strength will decrease significantly.

Therefore, when a steel structure or a reinforced concrete structure encounters a fire, the influence of the high temperature through the protective layer on the mechanical properties of the steel bar should be considered. Especially in the prestressed structure, the change of the entire structural stress system caused by the prestress loss of the steel bar under high temperature conditions must also be considered.

In view of the above reasons, precautionary measures should be taken in the steel structure, especially for high-rise buildings, including the installation of fireproof panels or painting of fireproof coatings. In reinforced concrete structures, the steel bars should have a protective layer of a certain thickness.

FIG.2-27 Apply fire-resistant paint

2.7.3 Corrosion and Corrosion Prevention Methods for Steel

Corrosion of steel refers to the damage of steel due to chemical or electrochemical action in long-term exposure to air or other media. Corrosion of steel is a serious problem. It not only reduces the effective cross-sectional area of the steel structure, but also causes the loss of the metal itself. More importantly, the metal structure and equipment suffer from the downtime and the production is reduced, and even

FIG.2-28 Corrosion of steel

accidents occur. Therefore, preventing corrosion of steel is of great significance to engineering structures.

1. Cause of corrosion

The main factors affecting the corrosion of steel are the ambient humidity, the type and quantity of aggressive media, the chemical composition of the steel and the surface condition.

According to the different effects of the steel surface and surrounding materials, the corrosion is generally divided into the following two types.

- Chemical corrosion

Chemical corrosion refers to the corrosion caused by the direct chemical action of steel and surrounding materials. As a result, iron oxide, iron sulfide, and the like are generated on the surface of the metal, and the metallic luster is reduced and the color is darkened. The degree of corrosion generally progresses slowly and increases with time, and the corrosion is accelerated in a humid environment and a high temperature.

- Electrochemical corrosion

Electrochemical corrosion refers to the corrosion that occurs after the steel and the electrolyte solution are in contact with each other to form a primary battery. Since there is a current generated in this process, a corrosion current is formed, which is called electrochemical corrosion, which is also called dissolved corrosion.

2. Method of preventing corrosion

- Protective film method

That is, a protective film is coated on the surface of the steel to be isolated from air or other

FIG.2-29 The anti-rust coatings

media, and neither oxidative corrosion nor electrochemical corrosion can occur. Commonly used protective layers are various anti-rust coatings (red dan+gray lead oil, alkyd enamel, epoxy zinc-rich, chlorosulfonated polyethylene anti-corrosion coating, etc.), enamel, paint, corrosion-resistant metal (lead, tin), Plastic or the like; or chemically treated to form an oxide film (blue treatment) or a phosphate film on the surface of the steel.

- Cathodic protection method

Cathodic protection is a method of protection based on electrochemical principles. There are two ways to implement this method:

a. Sacrificial anode protection method. That is, in the vicinity of the steel structure to be protected, especially underwater steel structures (such as ship shells, underground pipes, etc.), and more active metals such as zinc and magnesium, etc., so these more active metals in the medium The anode that becomes the corroded battery is corroded, and the steel structure is protected as a cathode.

b. Impressed current protection method. This method is to place some scrap steel or other insoluble metals in the vicinity of the steel structure, such as high-silicon iron and lead-silver alloy. The negative electrode of the external DC power source is connected to the protected steel structure, and the positive electrode is connected to the insoluble metal. After the energization, the insoluble metal is corroded as an anode, and the steel structure is protected as a cathode. Recently, measures to protect the protective film and the external power supply have been adopted, and the effect is better.

In addition, for the anti-corrosion measures of steel bars in reinforced concrete of port buildings, the cover layer may be laid with artificial rubber or polyvinyl chloride outside the concrete protective layer to avoid the penetration of sea water, and epoxy paint or asphalt paint may be used as the protective film. There is also a layer of zinc on the steel bar, which can improve the rust prevention ability by 5 to 6 times without reducing the grip strength of the concrete on the steel bar. In addition, some rust inhibitors (sodium nitrite, etc.) are infiltrated into the concrete, which can also delay the corrosion of the steel bars.

2.8　Asphalt

Asphalt is a derivative produced during the processing of crude oil. It is liquid, semi-solid or solid. It softens at high temperature, is brittle at low temperature, and has good cohesiveness and corrosion resistance. It is widely used as an anticorrosive material, bonding material and pavement material in civil engineering.

Since natural asphalt is too sensitive to temperature, the performance of asphalt should be

FIG.2-30　Asphalt as pavement material

improved according to engineering conditions and requirements, called modified asphalt. The modified asphalt refers to a mixture of a rubber, a resin, a high molecular polymer, a ground rubber powder, or a bitumen mixture which is subjected to mild oxidation processing of the asphalt to improve the performance of the asphalt. The modified asphalt improves the thermal stability of the asphalt, that is, the ability to resist rutting at high temperatures; reduces the cold and brittleness, that is, the ability to resist cracking at low temperatures; and enables the asphalt to be suitable for hot and low temperature environments. Some modified asphalts improve the adhesion of asphalt, improve its wear resistance, and greatly extend the life of asphalt pavement. The pavement paved with modified asphalt has good durability and wear resistance, which can ensure it does not soften under the high temperature and does not crack under the low temperature.

The advantages of modified asphalt are:

- high temperature resistance, low temperature resistance and strong adaptability;
- good toughness, anti-fatigue, and increase the bearing capacity of the road;
- resistant to water, oil and ultraviolet radiation, delaying aging;
- stable performance, long service life and reduced maintenance costs;
- reduce noise and improve driving comfort.

2.9　Emerging CE Materials

The polymer materials, new metal materials and various composite materials that emerged in the 20th century have fundamentally changed the function and appearance of civil engineering. On the one hand, the material properties are continuously improved and improved, and on the other hand, innovations are constantly being introduced. Lightweight concrete, high-strength concrete, high-performance concrete, fiber concrete, special concrete, high-strength steel and green building materials have entered the actual project from the laboratory.

The development of civil engineering materials can be summarized as:

- the application, scope and function of lightweight and high-strength materials are continuously expanding and improving;
- green building materials turn waste into treasure, recycling, and environmentally friendly;
- development of intelligent engineering materials;
- industrialization and productization of engineering materials production.

In civil engineering, the self-weight of building materials is one of the main loads in the project; reducing the weight of the material can reduce the structural load. Therefore, in recent years, a variety of lightweight engineering materials and lightweight components, such as ceramsite concrete, foam concrete, foam bricks, light steel components and inflatable membranes, have been introduced.

FIG.2-31　The foam concrete and inflatable membranes

1. New concrete materials

Previously high-strength concrete generally referred to concrete with a strength class above C45. With the development of science and technology, high-strength concrete is defined as concrete with a strength grade above C60. The highest strength concrete currently known for use in practical engineering is active powder admixture concrete, and the strength grade has reached C200.

Fiber concrete is a general term for a composite material composed of fiber and cement base (cement stone, mortar or concrete) by adding synthetic fiber or steel fiber to concrete. The main disadvantages of cement stone, mortar and concrete are: low tensile strength, low ultimate elongation, brittleness, and the addition of fibers with high tensile strength, high ultimate elongation and good alkali resistance can overcome these shortcomings. Fiber concrete can enhance the tensile strength of plastic concrete and significantly reduce its plastic flow and shrinkage microcracks. This ability to reduce or eliminate plastic cracks gives the concrete best long-term integrity. These fibers are evenly distributed throughout the concrete, providing additional reinforcement to the concrete and preventing shrinkage cracks. In concrete with fiber, it is also possible to minimize the width and length of cracks that may occur in concrete under stress.

2. Green materials

Green materials refer to materials that have minimal impact on the Earth's environment and are beneficial to human health in the processes of raw material adoption, manufacturing, use or recycling, and waste disposal. The basic characteristics of green building materials are: the use of natural resources as much as possible in building materials production, the use of tailings, waste

FIG.2-32 The anti-cracking fiber for concrete

residue, garbage and other wastes; the use of low-energy, pollution-free production technology; the use of formaldehyde, aromatics, hydrocarbons in production Compounds, etc., shall not use pigments, additives and products made of lead, cadmium, chromium and their compounds; the products not only do not harm human health, but also benefit human health; products have many functions, such as antibacterial, sterilization, mold removal, deodorization, heat insulation, fire prevention, temperature regulation, degaussing, radiation protection, anti-static and other functions; products can be recycled and recycled, non-polluting waste to prevent secondary pollution.

The development of green building materials in China has achieved initial results in recent years. For example, solid bricks, hollow bricks, blocks and other products developed from waste materials such as fly ash, coal gangue, slag and shale have replaced clay. Bricks have matured and have been widely used in engineering. The use of chemical gypsum board products, phosphorus gypsum, desulfurization gypsum, fluorogypsum instead of natural gypsum to produce "green building materials" can reduce the massive mining of natural gypsum deposits, and is one of the measures to solve the acid rain problem in some areas in China. Make full use of white polluting waste. After crushing, use the particleboard (or medium density board) production technology to produce insulation board series products and composite sandwich board products instead of wood. The HB color music board developed and produced by China Building Materials Research Institute belongs to this class. The straw or agricultural straw is used as the filler, and the cement grass is produced by autoclaving. The national industry standards for straw boards have been formulated and produced everywhere. The use of high-tech to produce multi-functional "green building materials" that are good for human health, such as antibacterial, mold-eliminating, deodorizing, sterilizing ceramic glass products, as well as adjustable wet, fire-proof, far-infrared inorganic interior wall coatings, do not emit organic volatiles waterborne coatings, non-toxic and highly effective binders. Cement production enterprises replace some cement with waste residue to reduce cement production; improve processes in production, reduce energy consumption, reduce pollutant discharge and carbon dioxide discharged into the atmosphere, and try to reach environmental tolerance. China's annual output of fly ash has exceeded 100 million tons, with an annual output of 80 million tons of slag, which is a reliable source of waste for the cement industry.

FIG.2-33 The fly ash autoclaved brick

3. Smart materials

Smart materials are a new class of composite materials that developed rapidly in the 1990s. Smart materials have the sensibility of sensing the environment (including the internal environment and the external environment), analyzing, processing, judging, and taking certain measures to respond appropriately to the intelligent features of the material. Among the civil engineering projects, shape memory alloys, piezoelectric ceramics, optical fibers, and electromagnetic currents have been put into use.

Shape memory alloys have a shape memory effect (FIG.2-34). The shape memory alloy can be used to make a fire alarm device and a security device for electrical equipment. When a fire occurs, the spring made of the memory alloy is deformed, and the fire alarm device is activated to achieve the purpose of the alarm. Another important property of shape memory alloys is pseudoelasticity, which is referred to as the large strain generated during loading that recovers with unloading. This property has been used in building shock absorption.

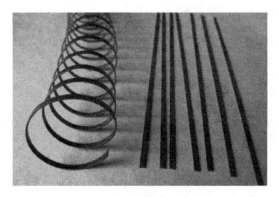 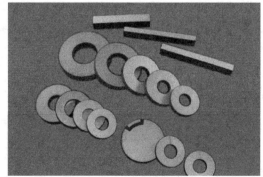

FIG.2-34 The shape memory alloys and piezoelectric ceramics

In line with the requirements of industrialization, engineering materials are also developing in the direction of commercialization and finalization, such as commercial concrete, prefabricated components, integrated dampers, and the like. Engineering materials are concerned not only with the materials themselves, but also with production, services and technology.

Questions

- How to classify architecture materials based on the role of materials?
- Briefly describe the development process of civil engineering materials.
- Try to list ten famous buildings, five of which are made of wood and five are made of stone.
- What are the characteristics of concrete and steel?
- What are the emerging civil engineering materials?

Reference List

[1]   Sharma, Bhavna Gatóo, Ana Bock, Maximilian Ramage, Michael. Engineered bamboo for structural applications[D]. University of Cambridge, Cambridge, UK, 2015.

[2]   "306R-16 Guide to Cold Weather Concreting". Archived from the original on 15 September 2017.

[3]   "Mapping of Excess Fuel Consumption". Archived from the original on 2 January 2015.

[4]   General Chipping. Confined Spaces: Understanding the Hazards of Concrete Chipping, General Chipping. Retrieved on November 5, 2018.

[5]   General Chipping. How Can I Encourage Cold Weather Safety for My Crew? General Chipping. Retrieved on November 5, 2018.

[6]   "Concrete in Practice: What, Why, and How?" (PDF). NRMCA-National Ready Mixed Concrete Association. Archived (PDF) from the original on 4 August 2012. Retrieved 10 January 2013.

CHAPTER 3
CONSTRUCTION ENGINEERING

Construction Engineering is the main part of civil engineering and the main content is the construction of various types of housing buildings, including various *industrial buildings* (FIG.3-1) and various *civilian buildings* (FIG.3-2). Construction Engineering should meet the basic requirements of practicality, beauty, economy and environmental protection. A building usually consists of three systems: architecture, structure and equipment. This chapter mainly introduces the content of the structure.

FIG.3-1 Industrial building   **FIG.3-2** Civilian building

Structure is the skeleton of a building, whose main function is to ensure the applicability, safety and durability of the building under various loads and other factors. The following briefly describes the *basic components* of the building structure and the main types of construction engineering.

3.1 Basic Components

The basic components of a building generally include foundations, columns, walls, plates, beams, arches, etc. As shown in FIG.3-3.

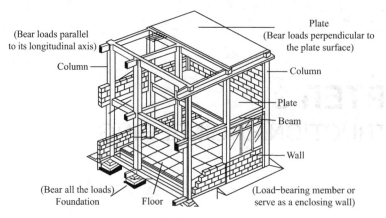

FIG.3-3   The basic components of a building

3.1.1   Foundations

All structures supported on rock and soil layers are composed of superstructure and foundation. The stratum bearing the building load is the groundsill, and the part between the superstructure and the groundsill is the foundation. As shown in FIG.3-4.

The *foundation* is the lowest component of the structure and is an important part of the building. The foundation bears all the loads transmitted from the upper structure of the building and transmits them to the groundsill along with its own weight. The vertical distance from the outdoor design ground to the bottom of the foundation is called *the embedded depth of the foundation*, as shown in FIG.3-5. The foundation is divided into shallow foundation and deep foundation according to the embedded depth of foundation.

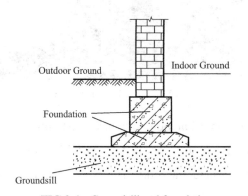

FIG.3-4   Groundsill and foundation

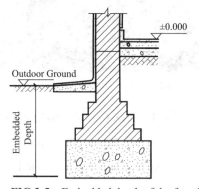

FIG.3-5   Embedded depth of the foundation

- *Shallow foundation*: independent foundation, strip foundation, cross foundation under column, raft foundation, box foundation
- *Deep foundation*: pile foundation, caisson foundation and deep foundation of underground continuous wall. As shown in FIG.3-6.

(a) Independent foundation

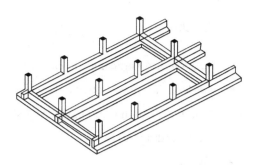

(b) Strip foundation

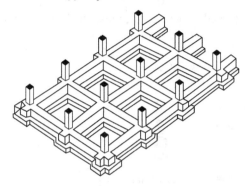

(c) Cross foundation under column

(d) Raft foundation

(e) Box foundation

(f) Pile foundation

(g) Caisson foundation

(h) Deep foundation of underground continuous wall

FIG.3-6　The type of foundation

3.1.2　Columns

- **Definition**: A vertical load-bearing member bearing loads parallel to its longitudinal axis.
- **Feature**: Its cross-sectional dimension is much smaller than its height.
- **Function**: Column in engineering structure mainly bears pressure and sometimes also bears bending moment.
- **Types**:

(1) According to the **materials** used: brick columns, block columns, wood columns, steel columns, reinforced concrete columns, steel reinforced concrete columns, steel tube concrete columns and various composite columns.

(2) According to the cross-sectional shapes (FIG.3-7 & FIG.3-8).

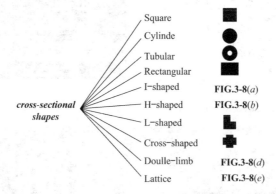

FIG.3-7　Sectional shape of the column

(a) I-shaped columns　　　　　　　　(b) H-shaped columns

(*c*) Cross-shaped columns

(*d*) Double-limb columns

(*e*) Lattice columns

FIG.3-8　The shapes of a columns

(3) According to the ***failure form*** or ***slenderness ratio*** of columns: short columns, long columns and medium-long columns.

(4) According to the ***stress forms*** of columns: axial compression columns and eccentric compression columns.

3.1.3　Walls

- ***Definition***: In general brick-concrete buildings, the wall is the main bearing component, and the weight of the wall accounts for 40 %~45 % of the total weight of the building. In other types of buildings, the wall may be a load-bearing member, or it may not be load-bearing but only serve as a enclosing wall (FIG.3-9).

- ***Types***: Walls can be divided into different types according to its position in the building, bearing mode and material (FIG.3-10 & FIG.3-11).

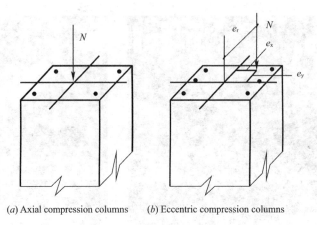

(a) Axial compression columns   (b) Eccentric compression columns

FIG.3-9   The force on the column

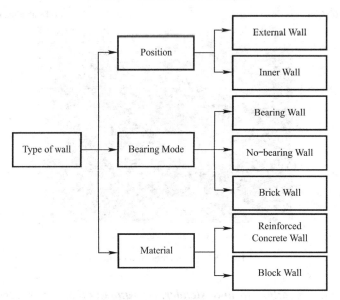

FIG.3-10   Types of Wall

- ***External wall***: The wall around the outside of the building is called the external wall. The external wall is the outer protective structure of the building and mainly plays the role of bearing load, wind prevention, rain protection, heat preservation and heat insulation.

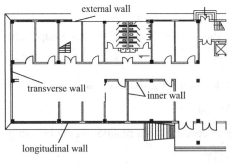

FIG.3-11   Types of Wall

- ***Inner wall***: The wall inside the building is called the inner wall and mainly serves as a load bearing and partition for the room. Among them, the walls arranged along the long axis of the building are called ***longitudinal walls***. The walls arranged along the short axis of the building are called ***transverse walls***.

- ***Bearing wall***: Walls directly bearing loads from floors and roofs are called bearing walls.
- ***Non-bearing wall***: Walls that do not bear these external loads are called ***non-bearing walls***. In the non-bearing wall, the wall that does not bear external load but only bears its own gravity and transmits it to the foundation is called the ***self-bearing wall***. The wall that only acts as a partition space and its own gravity is borne by the floor or beam is called a ***partition wall***.

3.1.4 Plates

- ***Definition***: A flexural member with a large plane size and a relatively small thickness (FIG.3-12).
- ***Feature***: Usually placed horizontally, but sometimes it can also be set obliquely (such as stair boards) or vertically (such as wall boards).

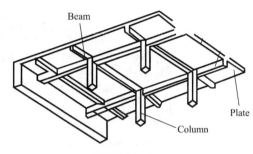

FIG.3-12 Plate

- ***Function***: The plate bears loads perpendicular to the plate surface, mainly bending moment, shear force and torque, but shear force and torque can often be ignored in structural calculation.
- ***Application***: Plate is generally used in floor, roof panel, balcony panel, stair panel, wall panel and so on.
- ***Requirement***: The plate needs to have enough strength and rigidity to ensure the safety and deformation requirements of the structure.
- ***Type***:

Plane form: square plate, rectangular plate, circular plate and triangular plate.

Section form: solid plate, hollow plate and trough plate.

Construction characteristics: precast slabs (factory prefabrication, site hoisting and laying), cast-in-place slabs (site integral casting).

Materials: boards, steel plates, reinforced concrete plates.

Stress form: one-way slab, two-way slab.

3.1.5 Beams

- ***Definition***: A flexural member bearing a load perpendicular to its longitudinal axis, and its cross-sectional dimension is much smaller than the span.

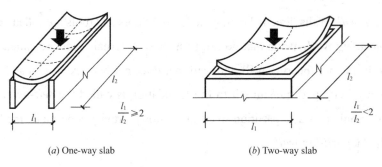

(*a*) One-way slab   (*b*) Two-way slab

FIG.3-13 One-way slab and two-way slab (Provided by sciencenet)

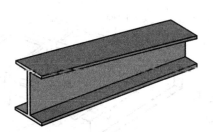

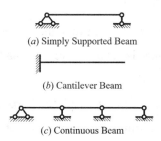

(*a*) Simply Supported Beam

(*b*) Cantilever Beam

(*c*) Continuous Beam

FIG.3-14 Variable cross-section beam   **FIG.3-15** Beam

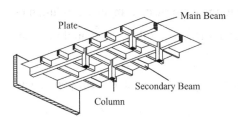

FIG.3-16 Main beam and secondary beam
(Provided by soyunpan)

- ***Feature***: The cross-section height of the beam is usually larger than the width of the cross-section, the ratio of the cross-section height to the span of the beam is generally 1/8~1/6.

- ***Function***: The beam bears the pressure and self-weight from the plate and transmits the force and self-weight to the vertical members such as columns or walls (FIG.3-17).

- ***Type***:

Section form: rectangular beam, T-beam, inverted T-beam, L-beam, box beam and so on.

Cross-sectional dimension: deep beam, flat beam, variable cross-section beam, stair beams.

Supporting mode: simply supported beam, cantilever beam and continuous beam.

Materials: steel beams, reinforced concrete beams, prestressed concrete beams, prestressed concrete beams, wood beams, steel and concrete composite beams, etc.

Position: main beam, secondary beam, ring beam, lintel, tie beam and crane beam.

3.1.6 Arches

- ***Definition***: a widely used structural form in building construction and bridge engineering.

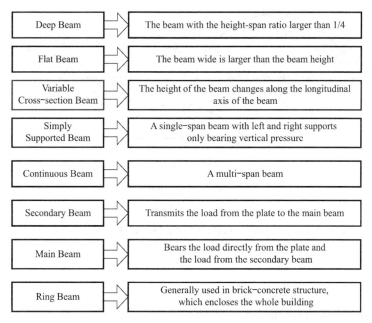

Deep Beam	The beam with the height-span ratio larger than 1/4
Flat Beam	The beam wide is larger than the beam height
Variable Cross-section Beam	The height of the beam changes along the longitudinal axis of the beam
Simply Supported Beam	A single-span beam with left and right supports only bearing vertical pressure
Continuous Beam	A multi-span beam
Secondary Beam	Transmits the load from the plate to the main beam
Main Beam	Bears the load directly from the plate and the load from the secondary beam
Ring Beam	Generally used in brick-concrete structure, which encloses the whole building

FIG.3-17   Definition or function of beams

Feature: under the action of load, the arch is mainly subjected to axial pressure and can be constructed from materials with poor tensile properties and strong compressive properties, such as brick, stone, concrete, etc.

FIG.3-18   Zhaozhou Bridge, located in Hebei Province   FIG.3-19   Hagia Sophia Church, located in Harbin

- *Application*: arch is mainly used as arched lintel for brick doors and windows in construction projects, and also for arched long-span structures.
- *Type*: according to the number of hinges, arches can be divided into ***three hinged arches, non-hinged arches, double hinged arches*** (FIG.3-20).
- *Main difference with beam*: the main internal force of the arch is axial pressure, while the bending moment and shear force are small or zero.
- *Function*: transmit and bear horizontal thrust.

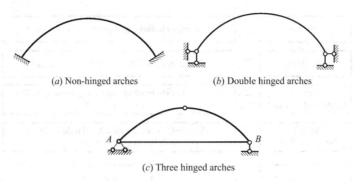

(a) Non-hinged arches   (b) Double hinged arches

(c) Three hinged arches

FIG.3-20   Type of arches

3.2   Truss Structure and Frame Structure

3.2.1   Truss Structure

- *Comprises*: Vertical bars, horizontal bars and diagonal bars.

In building construction, trusses are often used as roof bearing structures, which are often called roof trusses. There are two types of truss systems used for roof: plane truss, used for plane roof truss; Space truss, used for space truss.

- *Structural feature*: *Transform* the *whole bending* into the *compression* or *tension* of

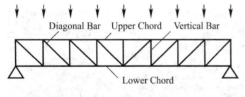

FIG.3-21   Truss Structure

local members. Truss structure (FIG.3-21) has been widely used in structural engineering due to its *reasonable stress*, *simple calculation*, *convenient construction*, *strong adaptability* and no lateral thrust to the support.

- *Disadvantage*: large *structure height* and small *lateral stiffness*. The height of the structure is large, which *increases the materials used* for roofing and retaining walls. The lateral stiffness is small, especially obvious for steel roof truss, the stability of the upper chord under compression is poor out of plane, and it is difficult to resist the longitudinal lateral force of the house, which requires support.

- *Loading feature*: The internal force of the structure is only axial force, without bending moment and shear force.

- *Calculation of structural internal force*:

Node method: Considering the balance of each node of the truss, the projection balance equation of each node is established step by step, and all unknown bar forces are obtained (FIG.3-22).

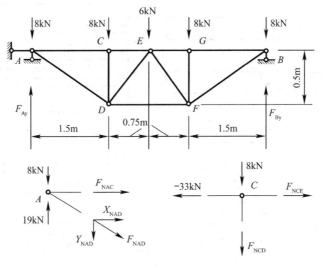

FIG.3-22 Node Method

Section method: Taking a truncated member as the local object of the truss, an equilibrium equation is established to obtain the required axial force of the member (FIG.3-23).

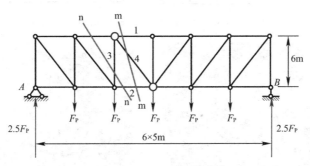

FIG.3-23 Section Method

- *Type*: Truss can be divided into different types according to its shape, structural stress characteristics and other factors, as shown in FIG.3-24.

Trusses are one of the common structural forms in the roof of large-span buildings. In general, when the span of a house is greater than 18m, trusses are more economical than beams. At present, the span of prestressed concrete roof truss in China has reached 60m, and that of steel roof truss has reached 70m (FIG.3-25 & FIG.3-26).

Roof frame that commonly used in our country contains triangle, rectangle, trapezoid, arch and do not have a variety of forms such as inclined abdomen pole roof frame.

3.2.2 Frame Structure

- *Definition*: a structure formed by connecting beam and column members through nodes.

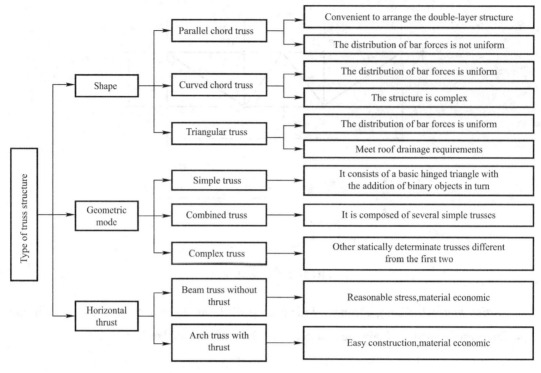

FIG.3-24   Type of truss structure

FIG.3-25   Triangle truss

FIG.3-26   Arch truss

FIG.3-27   Jiujiang Yangtze River Bridge

FIG.3-28   Tianjin Olympic center stadium

- *Advantage*: *flexible layout* of the building and can *obtain a larger space for use*.

In the frame structure, beams and columns replace load-bearing walls to release the space occupied by load-bearing walls, so the *flexible layout* of the building allows for *greater use of space*.

When needed, the partition can be divided into small rooms, or removed and changed into large rooms, so the use is *flexible*.

- *Disadvantage*: has *poor ability* to *bear horizontal loads*.

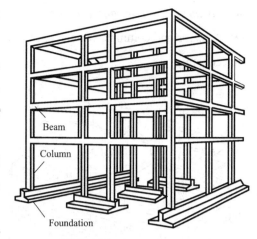

FIG.3-29   Frame Structure

The *lateral stiffness* of the frame structure is *small*, and the resistance to deformation under *horizontal action* is poor.

In order to meet the requirements of bearing capacity and lateral stiffness at the same time, the column *section* is often very *large*, very *uneconomical*, but reduce the use of area, so the frame structure in the earthquake area should not be too high.

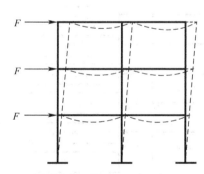

FIG.3-30   Horizontal action

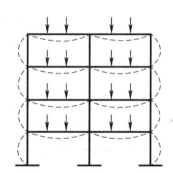

FIG.3-31   Vertical action

- *Application*: multi-story buildings and high-rise buildings with small height.
- *Mechanical behavior*:

The frame structure is a structural system that can *bear both vertical load and horizontal load*, because its stress system is composed of beams and columns, it is reasonable to bear vertical load, but the ability to *bear horizontal load is poor*, so this system is suitable for *multi-story buildings* and *high-rise buildings with small height*. China's code stipulates that the applicable height of reinforced concrete frame structures shall not exceed 12 stories.

The influence of *wind load* is small when *the number of floors of the building is small*, the *vertical load* plays a controlling role in the design of the structure, but *when the number of floors is large*, the *horizontal load* will have a great influence, resulting in a large cross-section size of beams and columns, which is not as reasonable as other structures in terms of technology and economy.

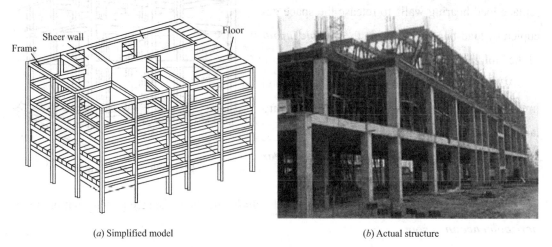

(*a*) Simplified model   (*b*) Actual structure

FIG.3-32   Frame structure (Provided by sina blog)

- *Type of frame structure*:

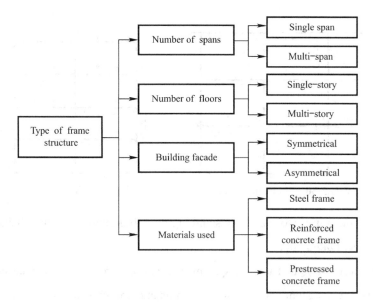

FIG.3-33   Type of frame structure

➢ *Requirements for frame structure*:

① The frame structure should be designed as *a bidirectional beam-column lateral resistance system*.

② It is not suitable to use *single span* frame in seismic design.

③ The **center line of beam and column** in frame structure should be coincident.

④ The **filling wall** of frame structure and **partition wall** APPROPRIATE chooses light weight wall.

⑤ When the frame structure is designed aseismatic, it should not adopt the mixed form which is partly supported by the **masonry wall**.

➢ **Difference between frame structure and brick concrete structure:**

✧ Bearing structure: The load-bearing structure of _frame structure_ is **beam, plate and column**, while the load-bearing structure of _brick concrete structure_ is **slab and wall**.

✧ Building storey: The height of _brick concrete structure_ cannot exceed **6 storeys**, while the height of _frame structure_ can be **dozens of storeys**.

✧ Effect of soundproof: The sound insulation effect of _brick concrete structure_ is **medium**, the sound insulation effect of _frame structure_ depends on **the choice of partition material**.

✧ Space transformation: Because most of the walls are not load-bearing, the _frame structure_ can be reconstructed by **removing the walls**. In the _brick concrete structure_, the wall is a load-bearing structure, so it is **not allowed to be removed**.

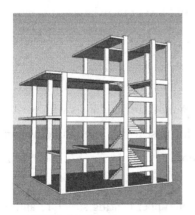
FIG.3-34 Frame structure

FIG.3-35 Frame-shear wall structure

➢ **Frame-shear wall structure:**

The frame-shear structure adds shear wall to the frame structure, which solves the problem that the vertical stiffness of the frame structure is not strong.

✧ Frame-shear structure is the combination of frame structure and shear wall structure, which can not only **provide a large space for the layout** of the building, but also have a **good lateral force resistance**.

✧ The shear walls in the frame-shear structure can be set separately, or the _elevator shaft,_

stairwell, *pipe shaft* and other walls can be used.

✧ The deformation of frame-shear structure is bending shear deformation, which reduces the relative displacement ratio between layers and the vertex displacement ratio, and *improves the lateral stiffness* of the structure.

✧ The horizontal load of the frame-shear structure is mainly borne by the shear wall, and the bending moments of the beams and columns are close to each other, which is conducive to *reducing the beam and column section*.

3.3 Single and Multi-story Buildings

3.3.1 Single story building

Single-story buildings include *general single-story buildings* and *large-span buildings*. General single-story buildings can be divided into *single-story civil buildings* and *single-story industrial plants* according to the purpose of use.

➤ **Single-story civil buildings**

Generally, *brick-concrete structure* is adopted, the material of the wall is brick and reinforced concrete slab or other materials are used for the roof. Most of them are used in single-story houses, public buildings, villas, etc.

✧ **Brick-concrete structure**

Brick-concrete structure is a structure with a small amount of reinforced concrete and a large part of brick walls. It is suitable for single-storey or multi-storey buildings with small space and small room area. Brick concrete structure is one of the most widely used building forms in China, but its shear strength under earthquake action is poor.

✧ **Masonry-timber structure**

The masonry-timber structure consists of a foundation, upper bearing structure and retaining structure. In the masonry-timber structure, the *walls* and *columns* of the vertical load-bearing structure are constructed by *brick or block masonry*, and the *floors* and *roof* frames are constructed by *wood*.

● *Feature*: The structure is simple, the materials are convenient and the cost is low, good practicability and durability. Usually used in rural areas, such as cottages, temple buildings, etc.

✧ **Bamboo structure**

● *Definition*: Bamboo structure is a building mainly constructed of bamboo materials.

● *Basic structural form*: Beams, trusses, arches, cage structures.

FIG.3-36   Brick concrete structure

FIG.3-37   Masonry-timber structure

FIG.3-38   Son La restaurant

FIG.3-39   The Nomadic Museum

- **Type**: Vertical, horizontal, roof structure, and space structure with the whole building as a whole.

- **Feature**: Comfortable living and convenient construction.

✧ **Other large span structures**

A large-span structure is a building with a span of more than **60**m. It is mainly used in public buildings such as **exhibition hall**, **gymnasium**, **city hall** and **airport terminal**.

FIG.3-40   Reticulated shell

FIG.3-41   Cable-membrane structure

(Provided by cn.made-in-China.com)

Its structure system commonly includes: *grid structure, reticulated shell, cable structure, cable-membrane structure*.

> **Single-story industrial plant**

The single-story industrial plant generally *adopts* reinforced concrete or steel structure columns, and the roof adopts steel roof truss structure.

According to the structural materials, it can be divided into *brick-concrete structure, steel structure, reinforced concrete structure, steel-concrete composite structure.*

According to the structural form, it can be divided into *bent structure* and *rigid structure*. *Bent structure* refers to the rigid connection between column and foundation, the connection between roof truss and column top is hinged, and *rigid structure* is also called frame structure, the connection between beam or roof truss and column is rigid connection.

| **FIG.3-42** Reinforced concrete structure | **FIG.3-43** Steel structure |

(Provided by gd.cnipai.com)

The bearing structure of single-story factory buildings in China mainly adopts *bent structure*. Most of these plants have *large spans*, *high heights* and *large crane tonnage*. This structure has *reasonable stress, flexible architectural design, convenient construction* and *high degree of industrialization*.

● *Load bearing member*:

✧ *Lateral bent*: foundation, column, roof truss.

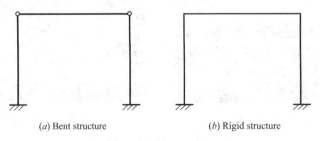

(*a*) Bent structure   (*b*) Rigid structure

FIG.3-44   Structure calculation diagram

✧ ***Longitudinal connection member***: Foundation beam, connection beam, ring beam, crane beam. Longitudinal connecting member and transverse frame constitute the framework to ensure the **integrity** and **stability** of the plant.

✧ ***Support system***: Roof bracing system and column bracing. The support system can ensure the **rigidity** of the plant.

● **Structural features:**

① The span is large, the height is big, bears the load big, the component internal force is big, the section size is big.

② The load forms are diverse, such as crane load, dynamic load and moving load.

③ The partition wall is less, the column is the main component to bears the load.

④ Large foundation force.

3.3.2 Multi-story building

● ***Definition***: Buildings with a height of ***more than 10 meters and less than 24 meters*** and with ***more than 3 floors and less than 7 floors***.

● ***Application***: residential buildings, shopping malls, office buildings, hotels and other buildings.

The basis for dividing the building according to the ***Design of civil buildings*** (GB 50352-2005) in China is as shown in Table 3-1.

Table 3-1　Type of building

Residential Building	floor	1~3	4~6		7~9		≥10
	type	low-story	multi-story		mid-highrise		high-rise
Other Building	height (miles)	≤24		>24		>100	
	type	single or multi story		high-rise		super high-rise	

Building height calculation

The height from the <u>outdoor design floor</u> of the building to its <u>eaves</u>, when *pitched roof* is adopted.

The height from the <u>outdoor design floor</u> to its <u>roof</u> surface, when *flat roof* is adopted.

When the same building has a variety of roof forms, the building height shall be calculated according to the above method and then the maximum value shall be taken.

A locally protruding roof is excluded from the height of the building.

● ***Advantage***:

a. Compared with the low-rise residential building, it has ***less land occupation*** and ***shorter construction period*** than the high-rise residential building.

b. There is ***no need*** to increase investment in ***elevators***, ***high-pressure pumps*** and ***public walkways*** as in high-rise residential buildings.

c. The ***structure design is mature*** and can be ***industrialized*** in large quantities.

d. The project *cost is low* and is *easily accepted* by buyers.

● *Disadvantage*:

a. The *living conditions* of the *bottom* and *top floors* are not ideal.

b. The *facade* and *style* of multi-story residential buildings are *inflexible* and lack of change.

FIG.3-45   Residential building

FIG.3-46   Shopping mall

● *Requirement of structure*: The overall *stiffness* and *seismic performance* of the super-structure of a multi-story building are very important, and the *selection of suitable foundation* has become an important factor affecting the safety and construction economy of the structure.

● *Basis of groundsill selection*:

✧ Requirements of designing institute, *engineering geology and hydrogeology report* of the proposed project site provided by the engineering exploration institute.

✧ The bearing capacity of *soft soil foundation* is low and the deformation of foundation is large, foundation treatment should be carried out.

✧ According to the *layout of the building, structure type, superstructure load size* and *distribution*, as well as the *seismic fortification zone* where the building is located, combined with the actual situation of the foundation of the building site.

✧ The site where the building is located is *adjacent to the building*.

● *Selection of foundation*:

✧ *Strip foundation under the wall.* Its main bearing compressive strength, but tensile, shear strength is not high. Generally, it is suitable for civil buildings with less than 5 floors and light production houses. This kind of foundation is characterized by being able to bear large load of superstructure and adapt to certain deformation of groundsill.

✧ *Independent foundation.* When the superstructure of a multi-story building is a frame system, if the bearing capacity of the foundation is high, the deformation of the foundation is small, and the distribution of load and column network is uniform, it is advisable to choose an independent foundation.

✧ *Cross foundation.* When the bearing capacity of the foundation is low and the column load is large, or when the deformation of the foundation and the distribution of the column load are not uniform in both directions, the cross foundation can be set up.

● *Common structural forms*: *brick-concrete* structure and *reinforced concrete* frame structure.

➢ **Multistory brick-concrete structure**

● *Definition*: A structural system supported by reinforced concrete floors and brick walls.

 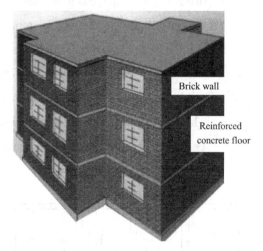

FIG.3-47   Multistory brick-concrete structure

● *Advantage*:

✧ The main load-bearing structure is made of bricks, which is easy to obtain materials.

✧ The walls are needed both for enclosures and partitions and as load-bearing structures.

✧ The vertical and horizontal wall layout of multi-story building can easily meet the structural requirements of the rigid scheme, so the rigidity of the masonry structure is relatively large.

✧ Construction is simple and fast.

● *Disadvantage*:

✧ The number of floors is limited, generally no more than 7 floors.

✧ The compressive strength is high, but the tensile strength and shear strength are low, so the seismic performance is poor.

✧ The space between the horizontal walls of multi-storey masonry buildings is limited, so it is impossible to obtain large space, so it is generally used for civil buildings and small or mediumsized industrial buildings.

- *Layout of wall*:

✧ **Transverse wall bearing system**

The load of the floor passes through the plate and beam to the *transverse wall*, which serves as the *main load-bearing vertical member*, while the longitudinal wall only serves as enclosure and separation.

✓ The transverse stiffness of the building is larger, the overall stiffness is good.

✓ The vertical wall is not a load-bearing wall can be opened larger doors and windows, facade designing is more convenient.

✗ Horizontal wall space is more dense, the flexibility of the room arragement is poor.

✧ **Longitudinal wall bearing system**

The load on the plate is transferred to the beam, and then from the beam to the *longitudinal wall*, which is the *main load-bearing wall*. The setting of *transverse wall* is mainly to meet the needs of building *stiffness and integrity*, its spacing is relatively large.

✓ The space of the room is bigger, planar arrangement is more flexible.

✗ The stiffness of the house is poor, and the vertical wall is stressed intensively.

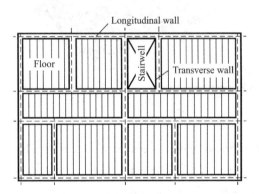
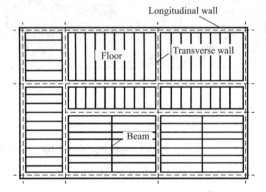

FIG.3-48 Transverse wall bearing system   **FIG.3-49** longitudinal wall bearing system

✧ **Transverse and longitudinal wall bearing system**

The plan that transverse and longitudinal wall bears load to be able to satisfy a room to width and depth.

The space between transverse walls is smaller than longitudinal wall bearing system, so the lateral stiffness of building is higher than longitudinal wall bearing system.

✧ **Internal frame bearing system**

On the basis of external wall bearing, some internal walls are replaced by reinforced concrete columns to obtain larger space.

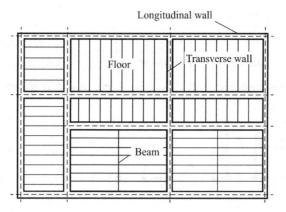

FIG.3-50 Transverse and longitudinal wall bearing system

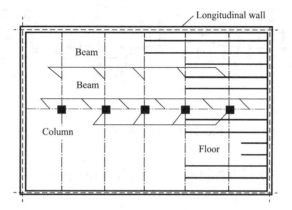

FIG.3-51 Internal frame bearing system

➤ **Multistory frame structure**

Frame structure can be divided into *steel frame* and *reinforced concrete frame* according to the different materials used.

According to different construction methods, reinforced concrete frame structure can be divided into *integral frame, semi-cast in site frame, prefabricated frame, prefabricated integral frame*.

FIG.3-52 Multi-story steel frame building

FIG.3-53 Prefabricated building

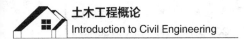

✧ **Arrangement of column grid**

The layout of the column grid should be considered from the ***following four aspects***: meeting the requirements of ***production technology***, meeting the requirements of ***building plane***, making the structure ***reasonable in stress*** and ***convenient in construction***.

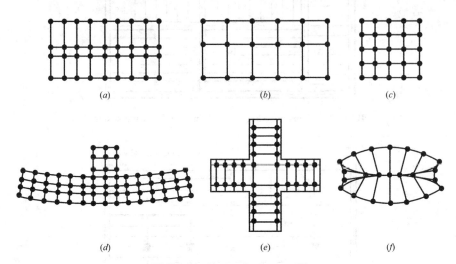

(a)   (b)   (c)

(d)   (e)   (f)

FIG.3-54 Form of column grid

3.4   High-rise and Ultra High-rise Buildings

3.4.1   High-rise Building

● ***Definition***: At present, there is no uniform standard for the definition of high-rise buildings in the world. Standards of some countries are shown in Table 3-2.

Table 3-2   Definition of high-rise building in different countries

Condition \ Country	America	Britain	Japan	China	
				Residential building	Other civil buildings
Height	>24.6m	≥24.3m	>31m	>28m	>24m
Floor	>7	—	>8	≥10	—

● ***History of development***:

High-rise buildings have been built since ***ancient times***. The Lighthouse of Alexandria, built by **Egypt** in 280 BC, is more than 100 meters high and constructed of stone. Built in 1056 in Yingxian County, **Shanxi Province**, the Sakyamuni Pagoda of Fogong Temple is more than 67 meters high and has a wooden structure.

FIG.3-55   The Lighthouse of Alexandria   **FIG.3-56**   The Sakyamuni Pagoda of
Fogong Temple

Modern high-rise buildings first emerged from the United States, and representative buildings are shown in FIG.3-57.

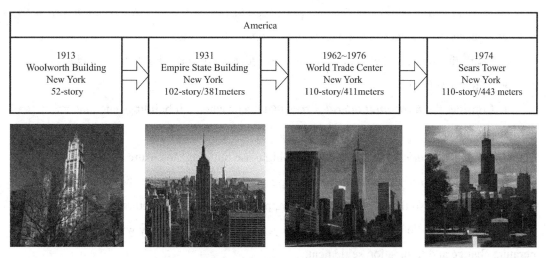

FIG.3-57   Representative high-rise buildings in America

China's modern high-rise buildings were built in the 1920s to 1930s, and the representative buildings are shown in FIG.3-58.

● ***Main point of design*:**

➢   Architectural Design

a. The ***general layout*** should increase the <u>fire separation distance</u>, and should deal with serious <u>sunshine interference</u>.

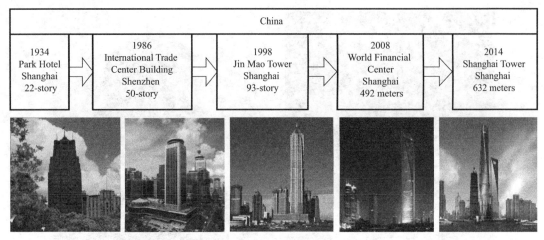

FIG.3-58 Representative high-rise buildings in China

b. On the basis of meeting the functional requirements, the repeated floor layout of multi-story buildings should be standardized to meet the requirements of ***vertical design techniques*** such as main structure, equipment pipelines and fire evacuation.

c. Reasonable arrangement of the ***number and layout of stairs and elevators*** to ensure efficiency and fire safety.

d. Internal and external construction practices must be adapted to ***deformation and safety problems*** caused by changes in wind, earthquake, temperature, etc.

e. In art, the image and effect of ***architectural form*** in the city should be considered.

➢ Structure Design

a. Consider the ***horizontal lateral forces*** that occur when tall buildings encounter wind and earthquake forces.

b. Control the ***ratio of height to width*** of high-rise buildings to ensure their stability.

c. The ***quality and stiffness*** of the building plan, shape and facade should be kept as symmetrical as possible, so that no weak parts can appear in the overall structure.

d. Properly handle the ***deformation of the joint structure*** caused by wind, earthquake, temperature change and foundation settlement.

e. Consider how to ensure ***safe and reliable design technology and construction conditions*** under heavy weight and deep foundation geological conditions.

- ***Structural forms***: frame structure, shear wall structure, frame-shear wall structure, tube structure, mega structure.

- ***Involving problem***:

a. With regard to the economic and environmental benefits of cities, they should be built in accordance with the sections and heights designated by the urban planning authority, rather than

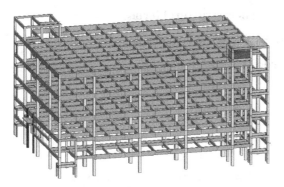

FIG.3-59 Frame Structure

FIG.3-60 Shear Wall Structure

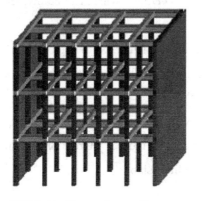

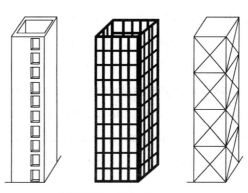

FIG.3-61 Frame-shear wall structure

FIG.3-62 Tube Structure

solely on the basis of the needs of the buildings themselves.

b. The construction difficulty of high-rise buildings increases, and the cost is higher than that of multi-story buildings. Therefore, all professional designers need to work together to reduce the cost.

c. The most important thing in high-rise building design is fire safety design. All professional designers should strictly abide by the regulations of high-rise building design fire code.

- ***Development trends*:**

a. Development and application of **new materials** and **high strength materials**. Now the strength grade of concrete has reached above C100, in order to achieve the purpose of lightweight and high strength, the development of *lightweight concrete* and *fiber concrete* has become a new trend. *Steel* is an ideal material for high-rise buildings. It will be a future trend to improve the strength, plasticity and weldability of steel.

b. **Lightweight** of building structure.

The increase of height is accompanied by the increase of building gravity, which increases the pressure on the vertical members and is bad for the seismic resistance. Therefore, it is of great significance to reduce the self-weight of buildings in terms of safety and economy.

c. The application of **new design concepts** and **new structural forms**.

The shape and structural system of modern buildings are complex and changeable, so new design concepts and structural techniques are needed.

3.4.2 Ultra High-rise Building

- *Definition*: a building with a height of *more than 100 meters* and a height of more than 40 floors.
- *Structural system*:

Ultra High-rise buildings are mostly *frame-core wall structures*, which can be divided into two types according to different heights:

The inner cylinder is a reinforced concrete core cylinder structure, while the outer cylinder is a giant column. The steel beams are connected between the giant column and the core cylinder, and the outer cylinder floor slab is a composite floor slab.

The inner cylinder is the reinforced concrete core cylinder, the outer cylinder is the giant column, the connection between the giant column and the core cylinder is the reinforced concrete beam, and the floor slab is the ordinary reinforced concrete floor slab.

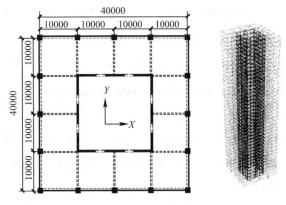

FIG.3-63 Frame-core wall structure

- *Advantage*:

a. The vertical and horizontal traffic inside the building can be used to shorten the distance so as to **improve the efficiency**.

b. The **land used** for large-area buildings has been greatly **reduced**.

c. **Reduce investment** in municipal construction and **shorten construction period**.

d. **Better ventilation** and **higher air quality**.

e. A **sign** of the degree of urban development.

- *Disadvantage*:

a. With regard to the economic and environmental benefits of the city, it should be built according to the location and control height designated by the city planning department, **not** according to the needs of the building itself.

b. The **cost is higher** than that of multi-story buildings.

c. **Fire safety design** should strictly comply with the provisions of the code for fire protection design of high-rise buildings.

FIG.3-64   Fire door   FIG.3-65   Fire separation distance

- *Construction Characteristic*:

a. The foundation adopts *deep foundation*.

b. The *basement* is high depth, much floors and large area.

c. The structure form is *mixed structure* mostly.

d. The *decoration engineering* quantity is large, the technical request is high.

e. Determine *safe and reliable design technology and construction conditions* under heavy weight and deep foundation geological conditions.

- *Examples of Ultra High-rise Building*: Top 10 ultra high-rise buildings in the world are shown below.

Note:

1. The year shown in the figure is the completion date of the building.

2. The number of building floors does not include the number of underground floors.

3. All pictures provided by www.nipic.com.

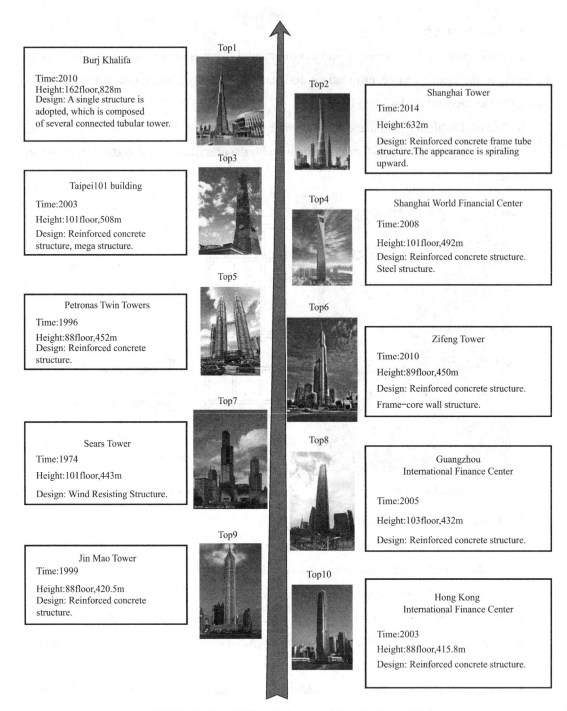

FIG.3-66 Top 10 Ultra high-rise buildings in the world

3.5 Special Structure

Special structure refers to engineering structure with **special purpose**, including high-rise structure, marine engineering structure, pipeline structure and container structure, etc.

3.5.1　Chimney

- ***Definition***: A high-rise structure for exhausting smoke into the sky.
- ***Feature***: Improve the combustion conditions and reduce the pollution of smoke to the environment.

At present, the tallest single tube reinforced concrete chimney in China is 210 meters. The tallest multi-flue reinforced concrete chimney is the 212-meter high four-cylinder chimney of ***Qinling power plant***. Dozens of chimneys over 300 metres high have been built in the world, such as the one at the ***Mitchell power station***, which is 368 metres high.

FIG.3-67　Qinling power plant

- ***Type***: Chimneys are divided into brick chimneys, reinforced concrete chimneys and steel chimneys according to materials.

Reinforced concrete chimney has small dead weight, beautiful shape, good integrity and seismic resistance, and easy construction, but the higher the chimney, the higher the cost.

The ***steel chimney*** has small dead weight and good anti-seismic performance. It is suitable for the site with poor foundation, but the steel chimney has poor corrosion resistance. It needs constant maintenance.

FIG.3-68　Reinforced Concrete Chimney

FIG.3-69　Steel Chimney

According to the structure system, the chimney can be divided into the following types, as shown in FIG.3-70.

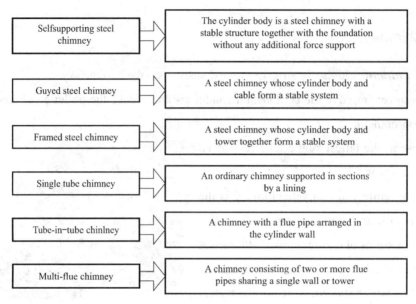

FIG.3-70   Type of chimney

3.5.2   Water Tower

- *Definition*: Water tower is a high-rise structure for **water storage** and **distribution**, and it is a common structure in water supply engineering.
- *Function*: maintain and regulate water quantity and water pressure in water supply network.
- *Component*: water tank, tower body and foundation.
- *Type*:

Material: reinforced concrete water tower, steel water tower, masonry tower body and reinforced concrete water tank.

Shape of water tank: cylindrical shell and inverted cone shell, spherical, box-shaped, bowl-shaped.

France has a **multifunctional water tower** with a water tank at the top, an office room in the middle and a shopping mall on the ground floor. China also has dual functional structures with chimneys and water towers built together.

3.5.3   Pool

- *Definition*: The pool is used **for water storage** just like a water tower.
- *Feature*: Water pools are mostly built on the ground and underground.
- *Fuction*: The clear water tank is used for *water supply* and the sewage tank is used for *sewage treatment*.

FIG.3-71 Inverted cone shell

FIG.3-72 Spherical

- *Type*: According to the <u>material</u> of the pool, there are *steel pool*, *reinforced concrete pool*, *steel mesh cement pool*, *brick pool*, etc. Among them, <u>the reinforced concrete pool</u> has the advantages of good durability, steel saving and simple structure, and is widely used.

According to the shape of the pool there are *circular pool* and *rectangular pool*. <u>Rectangular pool</u> construction is convenient and occupies less space. The force of the <u>circular pool</u> is reasonable and prestressed concrete can be used. Experience shows, small pool appropriate uses rectangle pool, the depth is shallow large pool also can use rectangle pool. Medium sized pool should be circular pool. Other forms of pools, such as circular sector pools, may also be used in consideration of topographical conditions. In order to save the use of land, also can use multistory pool.

FIG.3-73 Circular Pool

FIG.3-74 Rectangular Pool

3.5.4 Silo

- *Definition*: Silos are warehouses for storing bulk materials.
- *Type*:

<u>Use</u>: agricultural silos and industrial silos.

Agricultural silos are used to store grain, feed and other granular and powdery materials; **Industrial silos** are used to store coke, cement, salt, sugar and other bulk materials.

Material: reinforced concrete silo, steel silo and brick silo.

From the aspects of economy, durability and impact resistance, the most widely used in China is the monolithic cast ordinary reinforced concrete silo.

FIG.3-75   Industrial Silos   **FIG.3-76**   Agricultural Silos

The flat shape of the silo is *square*, *rectangle*, *polygon* and *circle*. Circular silo wall has reasonable stress, economical materials, so it is most widely used. When it is storage of a single type of material or small reserves, the layout should be a separate silo or single row. When the storage of material variety or large reserves, the layout should be group silos.

3.5.5   Power Transmission Tower

- *Definition*: The transmission tower is a high-rise structure used to erect conductors for power transmission in the air.

FIG.3-77   Transmission Tower

- **Comprise**: tower foundation, the tower body and the cross arm of the suspension wire.
- **Type**: According to the <u>shape</u>, there are two main categories: self-supporting tower and guyed tower. According to the <u>material</u>, it is divided into steel tower and reinforced concrete tower.

3.5.6　TV Tower

- **Definition**: A television tower is a building used for broadcasting and transmitting television signal.
- **Function**: TV tower can broadcast TV, also can go up to visit, has been combined with tourism, become a multi-purpose tower.

| **FIG.3-78**　Tokyo Sky Tree | **FIG.3-79**　Canton Tower |

3.6　Structure Calculation

In general, the ***purpose of structural design calculations*** is to ensure the safety, suitability and durability of the structure during construction and during use, and to make the structure most economical. ***The design calculation of structural members***, specifically refers to how to reasonably determine the ***cross-section*** and ***reinforcement*** of the members under certain conditions.

3.6.1　Calculation diagram

- ***Why should we simplify the actual structure***

Before the structural mechanics analysis, because the **actual structure is very complicated**, it is difficult and sometimes impossible to calculate completely according to the actual structure. Therefore, the actual structure should be simplified first, and a simplified figure should be used instead of the actual structure so that it can reflect the **actual main force characteristics**, while making the calculations greatly simplified.

- **Definition**: This simplified mechanical model used to replace the actual structure is called the calculation diagram of the structure.
- **Meaning**: The selection of the calculation sketch occupies a very important position in the mechanical analysis of the structure, which directly **affects the size of the computational workload** and the **difference between the analytical structure and the actual structure**.
- **Selection principle**:

(1) **Correctly reflect** the actual force of the structure, so the calculation results are as consistent as possible with the actual situation;

(2) The **secondary factors** that have less influence on the **internal force** and **deformation** of the structure can be greatly simplified or even neglected, which greatly simplifies the calculation.

- **The two aspects of the simplification of the structure**:

One is the simplification of the **structural system**; the other is the simplification of the **structural members**.

The **simplification of the structural system** means that the actual space system is simplified or decomposed into several **planar structural systems** under the possible conditions, so that the calculation of the entire space system can be simplified to the calculation of the planar architecture.

The **simplification of structural members** is mainly considered because the cross-sectional dimension of the rod is much smaller than its length. The cross-section stress can be calculated according to the internal force of the cross-section according to the plane assumption, and the internal force of the cross-section only changes along the length of the rod, so in the calculation diagram the **rod longitudinal axis can be used instead of the rod**, ignoring the effect of the cross-sectional shape and size.

3.6.2　Structural Design Information

Before we start designing structural components, we must have the following two aspects:

- **Component mechanics data**

The mechanical data mainly refers to the **internal force in the component under the load**, which is obtained by the mechanical calculation method according to the location of the component in the structure and the load combination.

Internal force is the basic basis for designing components. For example, to design a reinforced concrete beam:

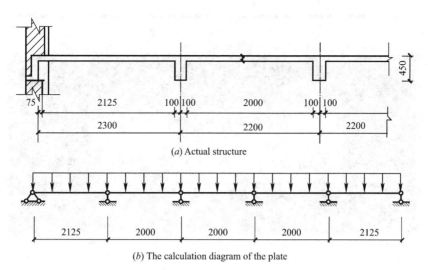

(a) Actual structure

(b) The calculation diagram of the plate

FIG.3-80　Calculation diagram (Provided by www.mfcad.com)

We need: the distribution of bending moments and the distribution of shear forces on the beam under the most unfavorable combination of loads.

We should: select the required section based on these internal forces and make reasonable design of the steel bar arrangement of the whole beam to ensure the bearing capacity and use requirements of the beam.

- ***Requirements for the use of components***

Carrying capacity: strength, stiffness, stability

Normal use requirements: crack resistance, crack width, deformation

3.6.3　Computer Structure Design Calculation

1. Structural Design Calculation Software

In order to simplify the work of structural engineers, software for integrated design, structural analysis, and construction drawing has been developed at home and abroad.

- ***Function***: **realizes** the structure arrangement, the force analysis of the whole structure, the force calculation of the components, and **automatically calculates the component reinforcement** according to the specifications of each country, and **checks the strength, stability and deformation** of the components, and can **calculate the analysis results** according to the calculation. Draw a **structural construction drawing**.
- ***Types***:

In China: PKPM series, Yingjianke, Guangsha, Lizheng, Bridge Doctor and other software.

In foreign countries: ETABS, STAAD, Pro, MIDAS, SAFE and other software.

(a)

(b)

FIG.3-81   Etabs interface (Provided by blog.zhulong.com)

What is PKPM?

PKPM is **a software system** for the **design**, **analysis**, **calculation and construction drawing** of frames, trusses, frame shears, shear wall structures and brick-concrete structures for reinforced concrete structures. It can be used in multi-story buildings or in high-rise buildings.

2. Structural Finite Element Analysis Software

The finite element method is a method based on mechanics theory, mathematical theory, and computer theory. With the rapid development of computer technology, software based on the principle of finite element method has emerged in the field of civil engineering. The **main finite element analysis softwares** include ANSYS, SAP2000, ADINA, ABAQUS, MIDAS, ALGOR and so on.

FIG.3-82 PKPM interface (Provided by www.newasp.net)

FIG.3-83 SAP2000 interface (Provided by www.leyijc.com)

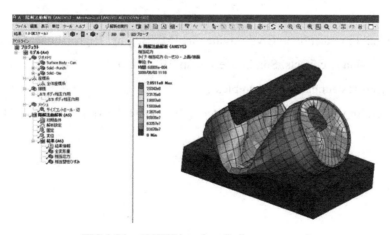

FIG.3-84 ANSYS interface (baike.sogou.com)

Questions

- What are the basic components of the building structure, and what are the functions of each component in the structure?
- What are the advantages and disadvantages of the truss structure and the frame structure?
- What are the types of single-story buildings? What are the definitions, advantages and disadvantages of multi-story buildings?
- What is the definition of high-rise buildings, and what are the advantages and disadvantages of super high-rise buildings?
- What are the special structures and what types of special structures?
- What principles should be followed in the selection of calculation sketches?

Reference List

[1]   Sun Airong, Liu Wancheng. English in Architectural Engineering[M]. Harbin: Harbin Institute of Technology Press, 1997.

[2]   Yao Yangping. English on Civil Engineering[M]. Beijing: Scientific and Technical Documentation Press, 1994.

[3]   Ren Jianxi. Introduction to Civil Engineering[M]. Beijing: China Machine Press, 2011.

[4]   Yan Xinghua, HuangXin. Introduction to Civil Engineering[M]. Beijing: China Communication Press, 2005.

[5]   Wen Tianxi. Introduction to Architectural Engineering[M]. Beijing: Chemical Industry Press, 2010.

[6]   Ye Zhiming. Introduction to Civil Engineering[M]. Beijing: Higher Education Press, 2016.

[7]   Vo Trong Nghia Architects. Son La Restaurant [EB/OL].[2017-10-12]. http://votrongnghia. com/projects/restaurant-son-la-complex/.

[8]   Pierre Frey. Simon velez: architect/mastering bamboo[M]. Arles: Actes Sud, 2013.

CHAPTER 4
CIVIL ENGINEERING STRUCTURES

Civil engineering is the general term for the construction of various engineering facilities.

It refers to the materials, equipment, survey, design, construction, maintenance, repair and other technical activities carried out, and also refers to the object of engineering construction. i.e. various engineering facilities constructed above ground or underground, on land or in water to directly or indirectly serve human life, production, military, scientific research, such as houses, roads, railways, pipelines, tunnels, bridges, canals, dams, ports, power stations, airports, marine platforms, water supply and drainage, and protection works, etc.

This chapter introduces road engineering, urban rail and underground railway, railway engineering, airport engineering, tunnel engineering, bridge engineering, development and utilization of subsurface, water resources and hydropower engineering.

4.1 Road Engineering

Roads are the basis of transportation, the basis of national economic activities and the infrastructure .With roads the population is floating frequently and our society more prosperous. The main function of the road is to serve as a link between the city and the city, the city and the countryside, the countryside and the countryside.

4.1.1 Road Construction

- **Geometric composition of the road**

Due to the limitations of natural conditions, the road has a turning point on the plane and undulations on the longitudinal plane. In order to meet the requirement of smooth, safe and speedy, the straight line on both sides of the turning point of the plane line and the slope point of the profile must be connected by a curve with a certain radius. The main content of highway graphic design is: according to the general direction of the route determined by the plan, under the premise of meeting the technical requirements of vehicle driving, combined with local terrain and address hydrological conditions, determine its specific direction; choose the appropriate curve radius to solve the connection at the turning point; ensure the driving line of sight, so that the route must meet the technical requirements, but also economic and reasonable.

The vertical section of the road is a longitudinal section of the route. It reflects the ground undulation of the midline and the slope of the design route. Since the terrain passing by the road is undulating, the car must travel on roads with different slopes, and the intersection of the longitudinal slope changes is called a grade change point. In order to facilitate driving, the curve set for mitigating the longitudinal slope line is called a vertical curve. According to the form of the slope transition, it is divided into a convex vertical curve and a concave vertical curve.

The cross section of the road is the normal direction of the road centerline, and is mainly composed by the side ditch, the shoulder and the driving part.

The intersection design of the route is also an important part of the road design. It is an important task to improve the capacity of intersections and reduce traffic accidents.

- **Road structure construction**

a. Subgrade construction

The subgrade is the basis of the driving part. It must have certain mechanical strength and stability, and it must be economical and reasonable to ensure the stability of the driving part and prevent natural damage. The cross section of a highway subgrade generally has three forms: embankment, cutting and part cut-part fill subgrade, as shown in FIG.4-1. The size of the subgrade consists of height, width and slope. The subgrade height is determined by the design of the longitudinal section of the route; the width of the subgrade is determined according to the designed traffic volume and the grade of the road; the subgrade side slope is determined by the overall stability of the subgrade.

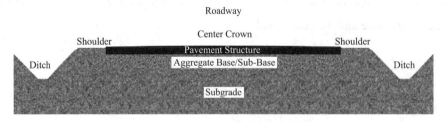

FIG.4-1　Subgrade structure diagram

b. Pavement construction

The road pavement is constructed of various hard materials layered for safe, rapid and comfortable driving. Therefore, the road surface must have sufficient mechanical strength and good stability, as well as surface smoothness and good anti-sliding properties. Pavement structure as shown in FIG.4-2.

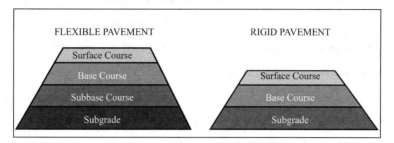

FIG.4-2   Road structure diagram

Pavement is generally divided into two types: *flexible pavement* and *rigid pavement*.

There is also a class of *semi-rigid materials*, mainly inorganic binders (cement, lime), hy-draulic materials (stabilized soil, sand, gravel) and industrial waste (such as fly ash, slag, etc.). The late strength of such materials increases greatly, and the final strength is higher than that of the flexible pavement, but lower than that of the rigid pavement. It is not wear resistant and is only used as a base layer for flexible or rigid pavements.

c. Drainage structure construction

In order to ensure the stability of the sub-grade and protection from surface water and groundwater, special drainage facilities should be built on the road.

The surface water removal system is divid-ed into *vertical drainage* and *lateral drainage* according to its drainage direction. Longitudinal drainage includes side ditch, intercepting ditch and drainage ditch. Horizontal drainage includes bridges, culverts, crown, ford, permeable embank-ments and aqueduct.

FIG.4-3   Expressway drainage ditch

d. Special structure of the road

The special structures of the road include tunnels, overhanging tables, anti-falling stone cor-ridors, retaining walls and protective works. The tunnel is a structure constructed for the passage of roads from interior the stratum or below the water layer. When the road passes over the moun-tains or through deep water, in order to improve the line shape of the flat and vertical planes and shorten the length of the route, tunnels are generally used to solve the problem.

e. Auxiliary structure and facilities

Generally, on the road, in addition to the above various basic structures, in order to ensure safe, rapid, comfortable and beautiful driving, it is also necessary to set up traffic management facilities, traffic safety facilities, service facilities and landscaping facilities.

Traffic safety management facilities are designed to ensure safe driving, traffic signs and pavement markings and traffic signals along the route.

Traffic signs are divided into three categories. Indicators: indicate the direction in which the driver is driving, mileage, etc.; warning signs: warnings of traffic obstacles and driving dangers ahead; prohibition signs: such as speed limit signs, load signs and signs of no parking.

Traffic safety facilities are designed to ensure the safety of the road and the role of the road. In the sharp bends and steep slopes of all levels of roads, it is necessary to set necessary safety settings, such as guard fence.

Service facilities generally refer to ferry terminals, bus stations, gas stations, repair stations, parking lots, restaurants, hotels, etc.

Landscaping facilities are an indispensable part of beautifying roads and protecting the environment.

4.1.2　Freeway

In order to meet the requirements of high traffic, high speed, heavy duty, safety and comfort of modern vehicles, highways have emerged. In recent years, many countries have built highways between major cities and industrial centers, forming a nationwide highway network. Some countries also lead major highways to other countries, known as international trunk highway as showed in FIG.4-4 and FIG.4-5.

FIG.4-4　Freeway in Houston Texas　　　　**FIG. 4-5**　Freeway in Los Angeles

- ***Linear Design Standards***

The geometric design standards of expressways are higher than those of other grades. The specific regulations vary from country to country.

The technical standards for highway engineering in China are mainly as follows:

a. Minimum radius of the horizontal curve and super high cross slope limit value

For a highway with a design speed of 120km/h, the general minimum radius of the flat curve is 1000m, the limited minimum of the horizontal curve is 650m, and the ultra-high cross slope is 10%.

b. Maximum longitudinal gradient and vertical curves

The maximum longitudinal slope curve of the expressway is 3% (plain terrain and rolling terrain) ~5% (mountainous terrain), the limited minimum of the vertical curve is 4000m (concave), 11000m (convex).

c. The requirement of the line shape

The expressway should ensure that the driver has a good line of sight, so there should be no sharp undulations and twisted lines, and the line shape should be continuous and smooth. that is, there should be no turning, misalignment, sudden change, void or cover in a certain line of sight. The lines have a good fit with each other.

d. Cross-section

There has to be at least two lanes in each direction of carriageway. Lane width 3.75m. Generally, the width of the median divider is 3.00m in the plain terrain and rolling terrain, the left marginal strip width is 0.75m, the central strip is 4.50m, and when the terrain is limited, the width will be 2.00m, 2.50m and 3.00m, respectively. In the hard road of the plain terrain and rolling terrain, the shoulder width is not less than 2.50m, and the width of the shoulder in the dirt road is not less than 0.75m, as showed in FIG.4-6.

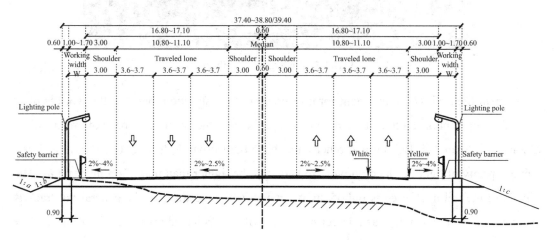

FIG.4-6 Typical cross section of 6 lanes divided freeway

4.2　Urban Rail and Underground Railway

4.2.1　Urban Rail and Underground Railway

At present, there are many cities with underground railways in the world, such as Paris in France, London in England, Moscow in Russia, New York and Chicago in the United States, Toronto in Canada, Beijing, Shanghai, Tianjin and Guangzhou in China.

Eight cities in the UK have subways, with a total length of nearly 1000 km, making them the longest subways of all countries. The subway in Moscow, Russia, is famous for its beautiful stations. The subway in New York is the most efficient subway system in the world. New York City currently has 26 subway lines, 468 subway stations, more than 6400 carriages, carrying 1.3 billion passengers a year.

FIG.4-7　Underground Railway

Urban light rail is an important form of urban rail construction, but also the world's most rapid development of rail transit form. In recent years, with the accelerated pace of urbanization, Chongqing, Shanghai, Beijing and other cities have built urban light rail. The weight and carrying capacity of light rail locomotives are known to be lighter than that of ordinary trains. The weight of the rail track used is only 50kg per meter, while the mass of the general rail track is 60kg per meter. It generally has a larger proportion of the dedicated track, mostly using shallow tunnels or viaducts, vehicles and communication signal equipment is also specialized, to overcome the tram running slowly, late, the shortcomings of loud noise.

- **Characteristics of urban light rail and subway generally:**

a. The control of noise and vibration is strict. Besides taking measures to reduce the vibration of the vehicle structure and building the sound barrier, the track structure is also required to take the corresponding measures.

b. Because of its high density, long running time and short working time, it is necessary to adopt strong track components and generally adopt track structure with less maintenance, such as concrete track bed.

c. In general, DC motor traction is used, and the track is used as the power supply circuit. In order to reduce the electrolytic corrosion of leakage current, high insulation between rail and foundation is required.

d. The curve segment occupies a large proportion, and the curve radius is much smaller than that of the conventional railway, generally about 100m. Therefore, the problem of curve track construction should be solved well.

FIG.4-8   Light Rail Mingzhu Line (Shanghai, China)

(https://media-cdn.tripadvisor.com/media/daodao/photo-s/0b/4f/3d/f5/caption.jpg)

4.2.2   Maglev railway

- *The origin of maglev railway*

The train running on the maglev railway is lifted up by the attraction and repulsive force generated by the electromagnetic system so that the whole train is suspended on the line and guided by the electromagnetic force. The DC motor is used to convert the electric energy directly into the propulsion force to push the train forward.

At present, there are three **types** of maglev in the world:

a. **The conventional magnet attraction suspension system** (EMS), represented by Germany, which makes use of the basic principle of the attraction between the conventional electromagnet and the ordinary iron-like material to attract the train and suspend the train; The suspension gap is small, generally about 10mm. Speeds up to 400km/h, 500km/h, suitable for long-distance rapid transportation between cities.

b. **The repulsive suspension system** (EDS), represented by Japan, which uses the superconducting magnetic levitation principle to create a repulsive force between the wheel and the rail, causing the train to run in the air. This maglev train has a large suspension gap, usually around 100mm. The speed can reach above 500km/h.

c. **China's permanent magnetic levitation**, which uses special permanent magnetic materials and requires no other power.

Compared with the traditional railway, maglev railway is suitable for high-speed operation because it eliminates the contact between wheel and rail, so there is no friction resistance and the vertical load of the line is small.

- *Development of maglev railway in various countries*

Research on maglev railways began in Japan in 1962. On July 31, 2003, the first test was carried out on the 500km/h train at the benefit Mountain in Yamanashi Prefecture, Japan.

Germany began to study maglev train in 1968. At present, Germany has a near maturity in the research of permanent magnetic levitation railway.

In 1989, the first maglev test train in China was built in Hunan University of National Defense.

FIG.4-9 Maglev Railway in Japan   **FIG.4-10** Maglev Railway in America

- *Challenges of maglev railway*

a. The maglev railway is very expensive. Compared with high-speed railway, the construction of maglev railway is expensive .

Germany believes that the maglev railway is much more expensive than the high-speed railway. According to a German estimate in the early 1980s, the cost of building a double-track maglev railway was about $6.59 million per kilometre. Paris to Lyon in France and Rome to Florence in Italy cost just $2.26million and $2.36 million per kilometer, respectively. The development of maglev railway is restricted by its large investment, long payback period and high investment risk coefficient.

b. The magnetic levitation railway cannot make use of the existing lines, and must be completely rebuilt.

Because the maglev railway is completely different from the conventional railway in principle and technology, it can not be used and reformed on the basis of the original equipment. The high-speed railway is different, through strengthening the roadbed, improving the line structure, reducing the curvature and slope and other aspects of the transformation, it is certain existing lines or certain sections to meet the high-speed railway traffic standards.

4.3   Railway Engineering

Before the invention of the railway, people used wagons as vehicles to carry goods and travel. But when a carriage travels in one place for a long time, the ground is muddy when it rains, and the wheels easily get stuck in the mud. So people want to put planks on the ground to make it easier for the wheels to travel. Later, in order to make the boardwalk can be used for a long time, the plank was covered with iron plate, which was the beginning of the railway track.

The greatest advantages of railway transportation are its large transport capacity, safety and reliability, high speed, low cost, little pollution to the environment, and little impact on the weather. Energy consumption is much lower than air and road transport, is the backbone of the modern transport system.

4.3.1   Railway Route selection Design and subgrade

- *Railway route selection design*

Railway alignment design is an overall work related to the overall situation in the whole railway engineering design. The ***main work contents of the route selection design*** are as follows:

a. According to the national political, economic and national defense needs, combined with the natural conditions, resource distribution, industrial and agricultural development of the areas through which the railway line passes, the basic trend of the railway line is planned, and the main technical standards of the railway are selected.

b. According to the terrain, geology, hydrology and other natural conditions along the line and villages, transportation, farmland, water conservancy facilities, to design the spatial location of the line.

c. Study the layout of various buildings on the line, such as stations, bridges, tunnels, culverts, roadbeds, retaining walls and so on, and determine its type and size, so that it is overall with each other, the overall economy is reasonable.

The design of the spatial position of the line includes **the design of the line plane** and **the vertical section**. Railway line plane refers to the projection of railway center line on horizontal plane, which consists of straight line section and curve section. The railway vertical section refers to the projection of the railway center line on the inside, which is composed of the slope section and the vertical curve connecting the adjacent slope sections. The characteristics of the slope segment are expressed by the length and gradient of the slope segment.

Railway alignment is to select the direction of the line on the topographic map or on the ground, and to determine the spatial location of the line. The basic methods of railway alignment are set line, spectacle line and spiral line, etc. Examples like FIG.4-11 and FIG.4-12.

 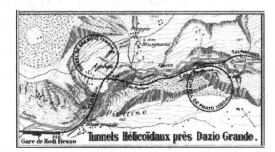

FIG.4-11　Spiral Line　　　　　　**FIG.4-12**　Spectacle-shaped railway line

- ● *Railway subgrade*

Railway subgrade is the structure of bearing and transferring the weight of the track and the dynamic action of the train, which is the basis of the track. The roadbed is a kind of earth-rock structure, because of all kinds of landforms, geology, hydrology and climate environment, what kind of disasters, such as flood, debris flow, earthquake and so on. Subgrade design generally needs to consider the following *issues*:

a. Cross-sectional form.

Embankment, half embankment, cutting, half cutting, version fill and dig, and so on, the same as the highway project.

b. subgrade stability.

The stability of roadbed may be affected by the train dynamic action and various natural forces, so the stability checking must be carried out in advance.

4.3.2   High-Speed Railways

An important symbol of railway modernization is to increase the running speed of trains by a large margin. High-speed railway is a kind of transportation means between cities in developed countries which developed gradually in the 1960s and 1970s. Generally speaking, ***the classification of railway speed*** is:

100~120km/h speed is called constant speed;

120~160km/h speed is called middle speed;

160~200km/h speed is called fast speed;

200~400km/h speed is called high speed;

400km/h speed above is called super high speed.

To sum up, there are ***several models*** for building high-speed railways in the world today:

a. Japan Shinkansen model: all new lines built, special for passenger trains (FIG.4-13).

b. German ICE model: all new lines built, mixed use of passenger and freight trains (FIG.4-14).

FIG.4-13   High-Speed Railways in Japan   **FIG.4-14**   High-Speed Railways in German

c. British APT model: neither new line, nor a large number of transformation of the old line, mainly by tilting the body of vehicles composed of EMU, passenger train and freight train mixed (FIG.4-15).

d. French TGV model: part of the new line, part of the old line transformation, passenger train dedicated (FIG.4-16).

Railway curve is one of the key factors to determine train speed. For example, curve speed limit is the first limit to speed increase on existing railways. For example, in China, the radius of the curve is usually 2000m for the railway of 160km/h per hour. The radius of the curve of the French railway, which reaches 300km per hour, is 4000m. Generally speaking, countries have their own technical standards and specifications for railway curves.

FIG.4-15 High-Speed Railways in Britain   **FIG.4-16** High-Speed Railways in France

The track ride comfort is an important problem to solve the train speed increase. The uneven track is the root cause of vehicle vibration and wheel-rail additional power. Therefore, the railway must strictly control the geometric state of the track in order to improve the smoothness of the track. At present, the high-speed railway track has realized the important of long track, which reduces the impact and vibration caused by the rail interface, and improves the ride comfort of the train.

In addition, the traction power of high-speed train is one of the important technologies to realize high-speed operation. It also involves many new technologies, such as: new power plant and transmission; traction power configuration; new train braking technology; high-speed electric traction technology; The structure of the car body and the moving part to meet the requirements of high-speed driving in order to reduce the air resistance of the new shape design.

The signal and control system of high-speed railway is the basic guarantee for the safe and high-density operation of high-speed train. It is an integrated control and management system which integrates microcomputer control and data transmission. It is also the latest comprehensive high technology for high-speed operation, control and management of railway, which is called Advanced Train Control Systems.

4.4 Airport Engineering

4.4.1 Airport profile

● *Civil transport aircraft and airports*

a. Overview of civil aviation transport aircraft

Trunk transport aircraft: refers to the carrying capacity of more than 100 people, longer than the range of 3000km transport aircraft. It is represented by Boeing757 (B757), B767, B747, B777 of American Boeing Company, A380 of European Airbus Company, Ir81 of Russia, etc.

Feeder transport aircraft: refers to the carrying capacity of less than 100 people, the range of 200~40km between the central city and small cities transport aircraft. To the United States DC3, British Aerospace Corporation SH330 and Bae146, China and the United States jointly manufactured MD82, MD90 (FIG.4-17) as a representative.

FIG.4-17　MD90

b. Overview of civil aviation airports

In the early days of civil aviation airport, the site was small and simple. In the period of civil aviation development (beginning in 1950), the airport has undergone qualitative changes, especially in the 1970s, the emergence of large wide-body passenger aircraft and the increase in traffic volume, so that the airport to the large-scale and modern forward. For example, Beijing Daxing International Airport was started constucting in 2014 and put into operation in Sept. 2019 (FIG.4-18 & FIG.4-19).

FIG.4-18　Beijing Daxing International Airport

FIG.4-19　Internal structure of Beijing Daxing International Airport

● *Classification and composition of civil aviation airports*

a. Airport classification

Airports are divided into international airports, trunk airports and feeder airports (FIG.4-20~FIG.4-23).

FIG.4-20   Shenzhen Bao'An International Airport

FIG.4-21   Los Angeles International Airport

FIG.4-22   Frankfurt airport

FIG.4-23   Sapporo New Chitose (CTS) airport

b. Composition of the airport

Civil aviation airports are mainly composed of flying areas, passenger terminal areas, cargo areas, aircraft maintenance facilities, fuel supply facilities, air traffic control facilities, security facilities, rescue and fire-fighting facilities, administrative office areas, living areas and auxiliary facilities.Construction, logistics support facilities, ground transportation facilities and airport air-space and other components.

4.4.2   Runway Program

Running to the airport is the theme of the flying area, directly for aircraft take-off taxiing and landing taxiing. When an airplane is taking off, it must now take off on the runway, taxiing and accelerating until the lift on the wings is greater than the weight of the aircraft before it can leave the ground. The plane landed at such a high speed that it had to slow down on the runway to stop. So the aircrafts are very dependent on the runway. If there is no runway, the aircraft on the ground can not fly, the aircraft can not land in the sky. Therefore, the runway is the most important airport engineering facilities.

Type: *The configuration of runway* can be divided into four basic forms: single runway, multiple runways, open V runway and crossing runway.

According to its *function*, runway can be divided into three kinds: main runway, auxiliary runway and take-off runway.

4.4.3 Aprons and Airport Clearance Areas

Airport aprons are mainly waiting pads and turn-around pads.

The former is used by aircraft for temporary parking while waiting for take-off or making way, usually on the side of a parallel taxiway near the end of the runway. The latter is used for aircraft to make a U-turn. When there is no taxiway in the flying area, a turn-around pad should be set up at the end of the runway.

Airport clearance area refers to the area involved in the take-off and landing of an aircraft. There should be an area around the airport where there are no obstacles affecting flight safety. In order to ensure the safety of aircraft, it is necessary to strictly control the height of the terrain and objects in this range and forbid any obstacles that may endanger the safety of flight. The regulations for the clearance area are controlled by such factors as aircraft take-off and landing performance, weather conditions, navigation equipment, flight procedures, and so on. ICAO specific requirements for airport clearance areas are shown in FIG.4-24.

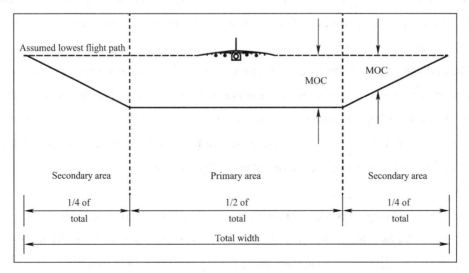

FIG.4-24   Required Obstacle Clearance

4.4.4　Layout of Terminal Areas

The planning and design of passenger terminal area is another important aspect of airport engineering. The passenger terminal area is mainly composed of the terminal, the platform and the parking space. The design of the terminal involves location, form, building area and other factors.

- **Terminal**

Terminal is the main building of the airport for passengers to complete from ground to air or from the air to the ground to change the mode of transportation. A terminal normally consists of the following five facilities:

a. facilities connected to ground transportation: kerbside roads for getting into and out of vehicles, bus stops, etc.

b. the facilities for handling all kinds of formalities: there are counters for passengers to check tickets, arrange seats, check luggage, as well as security checks, customs, border control counters and so on.

c. connecting aircraft facilities: waiting rooms, boarding facilities, etc.

d. airline operation and airport necessary management of office and equipment, etc.

e. service facilities: such as restaurants, shops, etc.

The layout of the terminal includes **vertical** and **horizontal layouts**.

The main consideration in the vertical layout of the terminal is to separate the passenger flow of departure and arrival passengers in order to facilitate passengers and improve operational efficiency, depending on the volume of passenger flow, the usable land area of the terminal building and the ground system, etc. The terminal may be arranged in the form of one, one and a half and two or more storeys.

The building area of the terminal is determined on the basis of peak hourly passenger traffic. The standard of area allocation is related to the nature, scale and economic conditions of the airport.

- **Terminal, airport car park and cargo area**

The apron, or aerodrome, is the apron located in front of the terminal. For aircraft parking, loading and unloading passengers, the completion of pre-departure preparations and post-arrival operations.

The airport parking lot is located near the terminal of the airport. Use a multi-storey garage when parking a lot of vehicles and land is tight. The construction area of the parking lot is mainly determined according to the peak hour traffic flow, parking ratio and the average area required per vehicle. The peak hour traffic volume can be determined according to the number of passengers, the number of visitors, the number of workers and clerks entering and leaving the airport, as well as the average number of passengers per vehicle.

Airport cargo area for cargo handling, loading aircraft and aircraft unloading, temporary storage, delivery and other uses, mainly by the business floor, cargo depot, loading and unloading yard and car park.

Cargo areas should be at a reasonable distance from passenger terminal areas and other buildings for future development.

4.5   Tunnel Engineering

A tunnel is a passage or space built under the ground, except that the aperture is too small, except for the so-called pipeline category. The OECD tunnel meeting defines tunnels as: caverns with a sectional area greater than $2m^2$ built by any method in the specified shape and size for any purpose. At present, the tunnel is still used for hydraulic tunnels such as railways, highway traffic, hydroelectric power generation, irrigation, etc. It can also be used for passages of large pipelines such as water pipes and transmission lines.

4.5.1   Highway Alignment

- *Highway tunnel lines*

The horizontal alignment of a highway tunnel is the same as that of a normal road, and is designed according to the requirements of highway regulations. The horizontal alignment of the tunnel is linear. Generally, the line is used to avoid the curve. If the curve must be set, the large radius curve should be used as much as possible to ensure the line of sight. The slope of the longitudinal section of a highway tunnel is determined by factors such as tunnel ventilation, drainage and construction. The slope is preferably gentle. The longitudinal slope of the tunnel should normally not be less than 0.3% and not more than 3%.

The space enclosed by the inner contour of the tunnel lining is called ***the tunnel clearance***. The tunnel clearance includes the boundary line of road-construction as shown in FIG.4-27.

Boundary lines are a kind of limit that no building such as tunnel linings can intrude. The boundary line of road-construction include the width of lanes, shoulders, curbs, sidewalks, etc; and the net height of lanes and sidewalks.

The cross-section clearance of highway tunnels, in addition to boundary line, includes the space required for auxiliary equipment such as pipelines, lighting, disaster prevention, monitoring, operation management, etc, as well as the margin and construction tolerances, as shown in FIG.4-28.

FIG.4-25 Tauern Railway Tunnel

FIG.4-26 Jarlsberg Tunnel

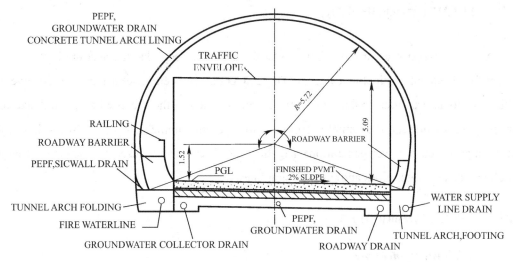

FIG.4-27 The Tunnel Clearance

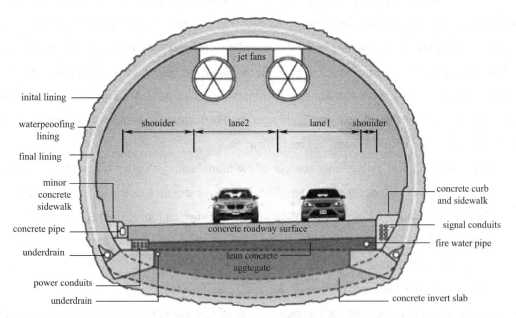

FIG.4-28 Auxiliary Equipments in the Tunnel

- *Road tunnel ventilation*

The car emits a variety of harmful substances, such as CO, NO, HC, SO_2, smoke and dust, causing air pollution in the tunnel. The main cause of damage caused by air pollution in highway tunnels is carbon monoxide. Fresh air can be introduced from the outside of the cave to dilute the concentration of carbon monoxide, and other gases can be at a safe concentration.

- *highway tunnel lighting*

Tunnel lighting is different from road lighting in general areas, and its distinctive feature is the need for lighting during the day. Prevent traffic drivers from causing traffic accidents due to insufficient visual information. It should be ensured that the bright and wide drivers who are accustomed to the outside world can still recognize the driving direction and drive normally after entering the tunnel. Tunnel lighting consists mainly of entrance lighting, basic non-illumination and exit lighting and continuous road lighting.

Entrance lighting refers to the visual illumination that must be guaranteed in order to adapt the driver from high brightness in the field to low brightness in the tunnel. It consists of three parts of the illumination of the critical part, the variable part and the mitigation part:

a. *The critical part* is a lighting measure taken to eliminate the black hole effect that the driver produces when approaching the tunnel.

b. *The variable part* is a section in which the brightness gradually decreases.

c. *The mitigation part* is the section where the driver enters the tunnel to the basic lighting of the habit, and adapts to the interval where the brightness gradually decreases.

Exit lighting is the illumination provided to prevent visual degradation when a car exits from a dark tunnel. The "white hole effect" should be eliminated.

- *Highway tunnel construction*

The main procedures for the construction of the tunnel main body are shown in FIG.4-29.

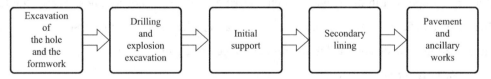

FIG.4-29 Construction schedule

Here are some main construction methods:

a. NATM

This is the abbreviation of the new Austrian tunneling method, abbreviated NATM.

The **concept** of NTAM was proposed by the Austrian scholar Professor L.V. Radcewicz in

1948. It is a method of combining the bolt and shotcrete as the main support means based on the existing tunnel engineering experience and the theoretical basis of rock mechanics. Later, this method has achieved extremely rapid development in many underground projects in Western Europe, Northern Europe, the United States, and Japan, and has become a method of constructing tunnels in soft and broken surrounding rock sections, with obvious technical and economic benefits.

b. NMT

The Norwegian Method of Tunnelling, referred to as NMT. It is a new method developed in the northwestern tunnel project in the 1990s. According to the Q value of the tunnel quality index, the method is to classify and select the surrounding rock, which is the perfection and supplementary development of NATM.

4.5.2　Railway tunnel

A railway tunnel is a building that is built underground or under water and is laid by a railway for locomotives and vehicles.

● **_Type_**: According to their location, they can be divided into three categories: **mountain tunnels** that cross the mountains or hills to shorten the distance and avoid large slopes; **underwater tunnels** that pass under the river or the seabed for crossing rivers or straits; **underground tunnels** that passes through the city to avoid affecting the normal operation of the city.

The underground railway is a complex underground engineering.

● **_Composition_**: Its composition includes facilities such as interval tunnels, subway stations and Interval equipment sections. Equipment used in underground railways involves unused technical fields.

The interval tunnel of the subway is the building connecting the adjacent stations. It occupies a large proportion in the length and engineering volume of subway lines. There should be sufficient space within the tunnel lining structure for vehicles to pass and lay tracks, power lines, communications and signals, cables and fire, drainage and lighting.

● **_Subway tunnel structure_**

a. Shallow-burying intersectional tunnel

Open excavation methods are used for shallow buried tunnels, and reinforced concrete rectangular frame structures are commonly used.

b. Deep-burying intersectional tunnel

Deep-buried tunnels are mostly used for mining method, with circular shield and reinforced concrete segment support. The depth of the overburden on the structure shall not be less than the diameter of the shield. From a technical and economic point of view, the construction of two single-track tunnels is more reasonable than the construction of a double-line tunnel in a large-section tunnel, because the single-track tunnel has a high utilization rate and is easy to construct.

The early underground railways in Moscow were adapted to the requirements of preparation for deep-buried forms, and some sections were as deep as 40~50m. Some of the London Underground is built in a clay layer about 30m deep.

- ***The form of the platform***

The platform is the most important part of the subway station. It is a venue for distributing passengers and passengers to board passengers.

The station platforms in different parts of the world have different forms. However, according to the positional relationship between it and the main line.

Type: It can be divided into: island platform, side platform and hybrid platform.

- ***Construction of subway tunnels***

The underground railway is arranged along the main street of the city and built in the urban area or on the outskirts of the city. Therefore, the selection of the construction plan should fully consider the impact of the subway on urban traffic, building demolition and the upper and lower pipelines, and consider technical and economic aspects. There are many ways to construct underground railways.

Type: There are two major construction methods, ***open excavation*** and ***mining method***.

FIG.4-30 Side Platform   **FIG.4-31**   Island Platform

➢ Open excavation

It is a method of constructing a tunnel by vertical excavation. It refers to excavation from the ground.

Construction Methods: The structure is constructed at the location where the underground railway structure is to be built, and then the construction method of backfilling the soil and restoring the road surface at the upper part of the structure. Or excavate from the ground, use large steel frame on both sides of the steel pile or diaphragm wall to maintain the original road traffic. The latter method is also called Cut and Cover, as shown in FIG.4-32.

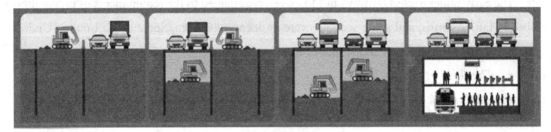

FIG.4-32 Cut and Cover

> Mining method

The mining method is the most commonly used excavation method.

Construction Methods: It is a construction method for excavating sections to build tunnels and underground works by drilling blasting methods, especially the excavation method of mining roadway technology used in hard rock layers. However, subway construction is mostly carried out in shallow soft soil layers, the main methods used are:

a. Shield method

The shaft is usually drilled at the location of the subway station or vent, and the shield is assembled in the well. From this position, the shield advances along the subway line to form a subway tunnel. The amount of earthwork it excavates, the migration of various underground pipelines and ground buildings and the traffic impact on the city are small.

The shield method is one of the tunnel excavation construction methods.

History: The shield method was used in the construction of the underground railway in 1874. At that time, a tunnel with an inner diameter of 3.12 m was constructed for the clay and water-bearing sandy ground on the east line of the London Underground Railway. The shield and the construction technique for grouting behind the lining were used. Since the 1940s, the Soviet Union has used underground shields with a diameter of 6.0~9.5m to construct underground railway tunnels in Moscow and Leningrad, and pushed the shield construction level to a new height. Since the 1960s, shield construction has been rapidly developed in Japan and has been widely used in underground railway construction in cities such as Tokyo, Osaka, Nagoya, and Kyoto.

The scope of application of shield method: Shield construction has the characteristics of fast construction speed, stable tunnel quality and little impact on surrounding buildings, and is suitable for construction in soft soil foundation.

Basic conditions for shield construction: In terms of line position, it is allowed to construct working wells for shield tunneling and slag feeding.

The tunnel should have sufficient depth of burial, and the depth of covering should not be less than the diameter of the shield.

Relatively homogeneous geological conditions.

If it is a single hole There is sufficient line spacing. The minimum thickness of the soil reinforcement between the hole and the hole and the hole and other buildings is 1.0m in the horizontal direction and 1.5m in the vertical direction.

From the economic point of view, the continuous construction length is not less than 300m.

Shield construction method: Construction preparation work including the construction of wells, shield base and support platform fabrication, installation and hole ground reinforcement. The basic schematic diagram of shield construction is shown in FIG.4-33.

FIG.4-33　Shield Construction

b. Grouting method

Place grouting holes in the construction area, pour cement mortar or other chemical slurry to consolidate the soil layer.

c. Immersed tube method

When the local railway is in the channel or river, the sinking method can be used. This is a major method of underwater tunnel construction.

The construction of the method is to prefabricate the prefabricated tunnel structure on the ship's platform or in the dock, and then transport the segment structure to the design position by floating or consignment in the water, and then ballast and sink it with water or sand. After the segment sinks to the pre-developed groove on the bottom of the water, the segment joint processing is performed. After all the segments have been connected, the trenches are backfilled to form an integrated tunnel.

d. Pipe jacking method

When the shallow subway tunnel crosses the ground railway, the urban traffic trunk line, the intersection, and the underground pipeline complex area, in order to ensure the traffic is not interrupted and the driving safety, the pipe jacking method can be used.

The pipe jacking method is to prefabricate the reinforced concrete tunnel structure in the working pit. After the strength is reached, the jack is used to push the structure to the design position. This construction technology is not only used for shallow subways, but also for urban water supply and drainage pipeline projects, urban road and ground railway intersections, and railway bridges and culverts.

4.5.3 Underwater tunnel

Feature: Compared with the bridge project, the underwater tunnel has the advantages of good concealment, smooth wartime, strong ability to resist natural disasters, and no obstacle to surface navigation, but its cost is high. The underwater tunnel can be used as a railway, highway, underground railway, shipping, pedestrian tunnel, or as a pipeline to the drainage pipe.

History: Since the 17th century, many canal tunnels have been built in Europe, of which the French Quedak Canal Tunnel is 157km long. In 1927, the Holland Tunnel was built at the bottom of the Hudson River in New York, and the next year, the world's first immersed tube tunnel, the Bosey Tunnel. At present, the longest railway tunnel in the world is the Japan Green Letter Undersea Tunnel passing through the Tsugaru Strait on the seabed. It has a total length of 53.85km and is constructed by mining method. In the soft soil, the underwater tunnel is mostly constructed by the immersed tube method and the shield method.

Advantages: It is generally considered that the immersed pipe method is cheaper, has a shorter construction period, and has better construction conditions, so it is more economical and reasonable.

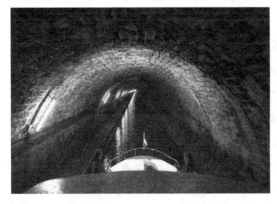

FIG.4-34 the French Quedak Canal Tunnel   **FIG.4-35** the Holland Tunnel

- ***The depth of the underwater tunnel***

The depth of the buried tunnel refers to the thickness of the soil layer of the tunnel under the riverbed. The depth is related to the length of the tunnel, the construction cost and the construction period. It is especially important that the thickness of the cover is related to the safety of underwater construction.

FIG.4-36   the Green Letter Undersea Tunnel

There are several major factors to consider when designing the depth of the underwater tunnel:

a. Geological and hydrological conditions

The geological characteristics of the riverbed through which the tunnel passes, the erosion and dredging of the riverbed.

b. Construction method requirements

Different tunnel construction methods have different requirements for the thickness of the top cover. According to the mining method construction, the empirical data of the buried depth depends on the strength of the surrounding rock to take 1.5 to 3 times the span of the hole. The immersed pipe method can be used as long as it meets the ship's anchoring requirements, about 1.5m. Shield construction, after years of research by international experts, the minimum cover thickness should be 1 times the diameter of the shield.

c. Anti-floating needs

Tunnels buried in quicksand and silt are subject to the buoyancy of groundwater. This buoyancy should be balanced by the weight of the tunnel and the weight of the soil covered by the upper part of the tunnel. For the sake of safety, the balance should be 1.10~1.15 times of the buoyancy.

d. Protection requirements

Underwater tunnels should have certain destructive capabilities against conventional weapons and nuclear weapons. The cover should have an appropriate thickness based on the requirements for indirect hits, loss reduction, and early nuclear radiation protection in conventional weapon attacks.

- ***Cross-sectional form of underwater tunnel***

a. Circular section

International subsea tunnels, especially riverbed sections, are mostly constructed by immersed tube method and shield method. The sections are mostly round, as shown in FIG.4-37 & FIG.4-38.

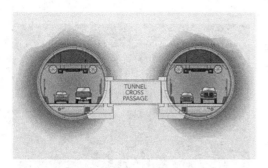

FIG.4-37   Tunnel Cross Section

FIG.4-38   Harbour Crossing of Hong Kong-Zhu-
hai-Macao Bridge

b. Arched section

When using the mining method, arched sections are generally used. In the form of arched sections, the force and section utilization are good.

c. Rectangular section

The Cannonier Underwater Tunnel in St. Petersburg, is a two-lane road tunnel with sidewalks and air ducts, constructed by immersed pipe method.

- ***Tunnel waterproof***

The main part of the underwater tunnel is in the rock layer under the river and seabed. Perennial below the groundwater level, it bears the full head pressure from the surface of the water to the depth of the tunnel. Therefore, the underwater tunnel has a waterproof problem from construction to operation.

The main measures for waterproofing are:

a. Waterproof concrete

The production of waterproof concrete mainly relies on adjusting the grading, increasing the amount of cement and increasing the sanding rate, so as to form a certain thickness of the wrapping layer around the coarse aggregate, and cut the passage of the capillary water seepage along the surface of the coarse aggregate to achieve the waterproof and water resistant effect.

b. Wall backfill

Backfilling behind the wall is to fill the gap between the tunnel and the surrounding rock to make the lining and the surrounding rock tightly combined, reduce the deformation of the surrounding rock, make the lining evenly pressed, and improve the waterproof ability of the lining.

c. Surrounding rock grouting

In order to improve the bearing capacity and reduce the water permeability of the surrounding rock of the underwater tunnel, pre-grouting can be carried out in the surrounding rock. In particular, the tunnel using the drilling blasting operation can grind the massive rock around the tunnel by grouting to form a certain thickness of the waterstop, and fill the cracks and cracks of the massive rock, thereby eliminating and reducing the water pressure on the lining

d. Double lining

The underwater tunnel is double lining for two purposes. One is the need for protection. Under the blast load, the surrounding rock may crack and break. As long as the lining waterproof layer is intact, there will not be a large amount of water in the tunnel and affect traffic. The second is to prevent high water pressure. Sometimes, although waterproof concrete is used for backfill grouting, it is inevitable that lining water seepage will occur under high water pressure. In this case, the double lining can be used as a waterproof measure for crossing the river section of the underwater tunnel.

- *Undersea tunnel*

The world's three major undersea tunnel project were:

a. **The Channel Tunnel** built in 1993 has a total length of 48.5km, a seabed section of 37.5km, and a tunnel with a maximum depth of 100m. The Channel Tunnel consists of two railway tunnels with an outer diameter of 8.6m and a service tunnel with an outer diameter of 5.6m. The completion of the Channel Tunnel is also the highest achievement in integrating the construction technology of advanced countries such as Britain, the United States, Japan and Germany (FIG.4-39).

FIG.4-39　The Channel Tunnel

b. **The Danish Dobel Tunnel** is a part of the trans-sea project. It is 7.9km long and consists of two railway tunnels with an outer diameter of 8.5m. The maximum depth of the tunnel is 75m. Because the tunnels that pass through the tunnel are ice and marl, they are all aquifers, and the amount of permeate is large, which is more difficult than the tunneling of the Channel Tunnel (FIG.4-40).

c. **Tokyo Bay Aqua Tunnel** in Japan is currently the longest submarine road tunnel in the world. It is 9.4km long and consists of two one-way highway tunnels with an outer diameter of 13.9m. The maximum depth is 50m. The shield design adopts the most advanced automatic tunneling management system, automatic measurement management system and automatic assembly system. The eight shields are docked on the seabed, which reflects the application of high technology in tunnel engineering. The Tokyo Bay Transected Highway Tunnel was completed and opened to traffic in 1998 (FIG.4-41).

FIG.4-40　The Danish Dobel Tunnel　　　　**FIG.4-41**　Tokyo Bay Aqua Tunnel

The United Kingdom and France signed an agreement in November 1986 to build the Channel Tunnel. The tunnel connects the city of Deauville with the city of Calais, France. When the tunnel is completed and opened to traffic, the time from London to Paris can be shortened by three hours, allowing the railway to compete with aviation. The width of the strait here is 36.8km and the length of the tunnel is about 51km.

4.6　Bridge Engineering

If there is no bridge, people must rely on ships to cross the river. When they encounter the canyon, they must detour, so they have to go a long way. Bridges are buildings built in human life and production activities to overcome the obstacles of natural gullies. They are also the oldest, most spectacular and most beautiful construction projects ever built by humans.

The bridge is not only a functional structure, but also a three-dimensional art project. It is also a landscape project with the characteristics of the times. The bridge has a magnificent glamour.

4.6.1　Type of bridge

According to the use of **nature**, it is divided into highway bridges, railway bridges, pedestrian bridges, machine-grown bridges, and overflow bridges.

According to **the size of the span** and **the total length of the multi-span**, it is divided into small bridge, middle bridge, ordinary bridge and special bridge. As shown in the table below.

Table 4-1　Bridge types

Type of bridge	Length of Multi-span (L/m)	Length of single span (L_0/m)
Special bridge	$L \geqslant 500$	$L_0 \geqslant 100$
Ordinary bridge	$L \geqslant 100$	$L_0 \geqslant 40$
Middle bridge	$30 < L < 100$	$20 \leqslant L_0 < 40$
Small bridge	$8 \leqslant L \leqslant 300$	$5 < L < 20$
Culvert	$L < 8$	$L_0 < 5$

According to the position of the carriageway, it is divided into the deck bridge, the half-through bridge and the bottom-through bridge.

According to the bearing condition of the load-bearing members, it can be divided into beam bridge, slab bridge, arch bridge, steel bridge, suspension bridge and combined system bridge.

According to the age of use, it can be divided into permanent bridges, semi-permanent bridges and temporary bridges.

According to the type of material, it is divided into wooden bridge, masonry bridge, reinforced concrete bridge, prestressed bridge and steel bridge.

4.6.2　Master plan and design points of bridge engineering

1. Tasks and priorities of the bridge master plan

The basic contents of the bridge master plan include: selection of bridge position; determination of total span and hole splitting scheme of the bridge; selection of bridge type; determination of longitudinal and cross-section layout of the bridge. The principle of bridge master planning is to fully implement the guidelines of safety, economy, application and aesthetics according to the tasks, nature and future development needs. The following requirements need to consider:

- *Requirements for use*

The carriageways and sidewalks on the bridge should ensure the safe and smooth flow of vehicles and pedestrians to meet the needs of future traffic development. Bridge type, span size and clearance under the bridge should also meet the requirements of flood discharge and safe navigation.

- *Economic requirements*

The construction of the bridge should reflect economic rationality. The selection of the bridge plan should fully consider the physical conditions such as local conditions and local materials and construction level, and strive to minimize the total cost and material consumption and

the shortest construction period on the basis of meeting the functional requirements.

- ***Structural requirements***

The entire bridge structure and its components shall have sufficient strength, rigidity, stability and durability during manufacture, transportation, installation and use.

- ***Aesthetic and environmental requirements***

The bridge should have a beautiful shape and should be coordinated with the surrounding environment and scenery. At the same time, it must also reflect the concept of environmental protection, and it will not affect or destroy the surrounding environment and ecology.

2. Key points of bridge engineering design

- ***Select the bridge position***

Under the premise of obeying the general direction of the route, the bridge is selected in the river section where the river is straight, the river bed is stable, the water surface is narrow, and the water flow is stable.

- ***Determine the total span and bore number in the bridge***

The length of the total span should ensure that there is enough cross-section under the bridge to smoothly vent the flood and pass the ice. According to the geological conditions of the riverbed, the allowable scouring depth is determined to properly compress the total span length and save costs. The bore number and the size of the span should take into account the navigation requirements of the bridge, the advantages and disadvantages of the engineering geological conditions, and the high and low cost of the project. Generally, the larger the span, the larger the total cost and the more difficult the construction. The bridge elevation is determined while determining the total span and bore number. The design of navigable water level and navigation clearance height is the main factor determining the elevation of the bridge. Generally, under the premise of satisfying these conditions, the value should be taken as low as possible to save the project cost.

- ***Vertical and horizontal section layout of the bridge***

The longitudinal section of the bridge is to consider the connection between the road and the bridge after determining the total span of the bridge and the elevation of the bridge. The connection line shape should generally be based on the terrain and line requirements of the bridge ends at both ends. The longitudinal slope is for bridge deck drainage and is generally controlled at 3%~5%. The cross-section arrangement of the bridge includes the width of the bridge deck, the lateral slope, and the arrangement of the bridge span structure. The width of the bridge deck includes the width and structural dimensions of the roadway and sidewalk. Different countries have different regulations according to the road grade.

- ***Road bridge type selection***

Bridge type selection refers to what type of bridge is selected, whether it is a beam bridge or an arch bridge; whether it is a steel bridge or a cable-stayed bridge; a porous bridge or a single-span bridge. Generally, comprehensive consideration should be given from the aspects of safety, practicality and economic rationality, and the optimal bridge type scheme should be selected. In actual operation, it is often necessary to prepare multiple sets of possible bridge types. After comprehensive comparative analysis, it is necessary to find the optimal solution that meets the requirements.

4.6.3 Structural form of the bridge

1. Beam bridge

The beam bridge is the most common and basic bridge.

Feature: The beam-type bridge power transmission is that the load of the upper structure of the bridge is transmitted vertically to the support and then to the lower structure. The bridge between the two supports must bear a large bending moment force.

The discontinuity on the pier is called a **simply supported beam**.

A continuous beam on a pier is called a **continuous beam**.

A simple bracket supported on a cantilever is called a **hanging beam**.

A beam with a cantilever extending is called an **anchor beam**.

At present, the most widely used is the bridge in the form of simply supported beam structure. This type of structure is simple, the construction is convenient, and the requirements for foundation bearing are not high, and the bridges with a span of 25m or less are usually adopted. When the span is greater than 25m and less than 50m, the form of prestressed concrete simply supported beam bridge is generally used.

(*a*) Lagentium Viaduct, Castleford   (*b*) Elbe bridge Bad Schandau

FIG.4-42 Beam Bridge

Forces acting on the structure

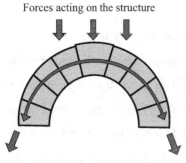

FIG.4-43   the Force of Arch Bridge

2. Arch bridge

Feature: Unlike a beam bridge, an arch bridge is subjected to its arched oblique compression force rather than bending force. The arch bridge is composed of arch rings or arch ribs as the main load-bearing structure. Under this vertical load, the pier or abutment will be subjected to horizontal thrust, and its force is shown in FIG.4-43. The bending moment and deformation of the arch are relatively small, mainly subject to pressure, so the arch bridge is constructed with brick, stone, concrete and reinforced concrete materials.

The arch bridge has a large spanning capacity and a beautiful appearance, so it is generally economically reasonable to construct the arch bridge. However, due to the large horizontal thrust at the pier or abutment, the lower structure of the bridge and the requirements for the foundation are relatively high. In addition, the construction of arch bridges is more difficult than beam bridges.

3. Steel Bridge

The standard beam bridge, the structure of the bridge's girders and piers is separate. The shape of the steel bridge is similar to that of a beam bridge. However, the upper structure of the steel bridge and the lower leg portion are completely rigid. A steel bridge is a bridge structure in which a beam and a column (or a vertical wall) are integrally combined (FIG.4-44).

Form of Force: Under the vertical dynamic load, the beam is mainly bent, and there is horizontal thrust at the foot of the column. The force state is between the beam bridge and the arch bridge.

FIG.4-44   Steel Bridge in New England

FIG.4-45   Bridge Design and Construction

Types: Steel bridges can generally be used in three types: T-shaped steel bridges, continuous steel bridges, and diagonal steel bridges. The T-shaped steel frame is easy to apply pre-stress, and can be made into a large-span steel frame after adding the hanging beams at the two end ends, which are often used in large-span bridges that cross deep water, deep valleys and large rivers.

Feature: For the same span, under the same external force, the cross-positive bending moment of the steel bridge is smaller than that of the ordinary beam bridge. The height of the building in the steel bridge span can be made smaller. When encountering a three-dimensional crossover line in a city or when it is required to cross a river, the bridge type can minimize the line elevation and improve the longitudinal slope of the bridge. When the deck elevation is determined, it can increase the clearance under the bridge.

For steel bridges, prestressed concrete structures are usually used.

4. Cable-stayed Bridge

The beam is tied to the tower with a number of diagonal cables to form a cable stayed bridge. In contrast to the porous beam bridge, a diagonal cable replaces the fulcrum of a pier, thereby increasing the span of the bridge.

In the middle of the 20th century, due to the emergence of computers, the problem of cable calculations was solved. The improvement of the adjustment device solves the control problem of the cable, which makes the cable-stayed bridge the fastest growing bridge type in the past 50 years.

Contents: The cables-tayed bridge is composed of three basic members: main beam, tower column and diagonal cable.

Feature: The slings made of high-strength steel hoist the main beam at multiple points, transfer the dead load and vehicle load of the main beam to the tower column, and then pass it to the foundation. It is a structure in which the bridge deck system is pressed and the support system is pulled. In this way, the cable can make full use of the tensile properties of the high-strength steel, and can significantly reduce the cross-sectional area of the main beam, and the structure is greatly reduced, so that a large-span bridge can be constructed.

The main beam and tower of the cable-stayed bridge can be constructed of reinforced concrete or section steel.

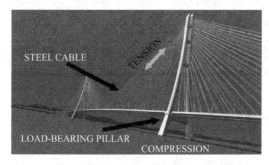

FIG.4-46   the Function of Cable-stayed Bridge

FIG.4-47   Cable-stayed Bridge

Cable-stayed bridges can be built in different types of single towers, twin towers or multiple towers depending on the size of the span and economic considerations. Generally, when the symmetrical section and the clearance requirement under the bridge are large, a double tower type is often used.

Advantages: The cable-stayed bridge is the most imaginative and connotative bridge type since the 1950s, and it has a wide range of adaptability. As a cable system, cable-stayed bridges have greater spanning capability than beam bridges. Due to the anchoring characteristics of the cable, there is no need for a huge anchor like a suspension bridge. The cable-stayed bridge has good mechanical properties and economic indicators, and has become the most important bridge type for long-span bridges.

Problems: It must be noted here that the cable of the cable-stayed bridge is the lifeline of the bridge. So far, there have been several unfortunate engineering examples in the world that have been completely changed due to severe corrosion of the slings in the past few years. Therefore, how to protect the cable and ensure its service life is still a concern of today's bridge engineering.

5. Suspension Bridge

Definition: The bridge on which the deck is supported on the suspension cable is called a suspension bridge.

Feature: In contrast to the arch ribs, the cross section of the cable can only withstand tensile forces. The simple suspension bridge for people and animals often lays the bridge deck directly on the suspension cable. It is not possible to use a suspension bridge with modern means of transportation. In order to maintain a certain degree of straightness of the bridge deck, the bridge deck is hung on the suspension cable with a sling. Unlike the arch bridge, the arch rib as a load-bearing structure is rigid, and the sling as a load-bearing structure is flexible. In order to avoid the deformation of the bridge deck with the suspension cable when the vehicle is running, the modern suspension bridge is generally provided with a rigid beam.

Form of Force: The bridge deck is laid on a rigid beam and the rigid beam is suspended from the suspension cable. The vertical moving load acting on the bridge surface can be loaded into the main tower through the main beam and the boom and transmitted to the foundation. During the transmission of force, both the boom and the main cable are subjected to a large pulling force, which is balanced by the giant anchors built after the bridges on both sides of the bridge.

Advantages: Suspension bridge is one of the main forms of extra-large span bridges. Due to the beautiful shape and grand scale of suspension bridges, people often call it the "Queen of Bridges". The suspension bridge can make full use of the strength of the material, and has the characteristics of saving materials and light weight. Therefore, the suspension bridge has the largest spanning ability among various system bridges, and the span can reach more than 1000m. When the span is greater than 800m, the suspension bridge scheme is very competitive.

For example: The suspension bridge consists of four parts: the main cable, the tower, the stiffening beam and the anchor. The main slab is manufactured by the AS method or the PPWS method. The United States, the United Kingdom, France, Denmark and other countries all adopt the AS method, and China and Japan adopt the PPWS method. The tower form generally adopts a door type E frame, the material is made of steel and concrete, the United States, Japan, and the United Kingdom use more steel towers, and China, France, Denmark, and Sweden use concrete towers. The stiffening beam has steel truss beams and flat steel box girder. There are more steel trusses in the United States and Japan, and more steel boxes are used in China, Britain, France and Denmark. There are gravity anchors and tunnel anchors in the anchors, and more gravity anchors are used.

History: The modern suspension bridge began in 1883 with the completion of the Brooklyn Bridge in the United States and has a history of more than 120 years. In the 1930s, the successively built George Washington Bridge and the Golden Gate Bridge in San Francisco made the suspension bridge span more than 1000m. In 1964, the British Saiwen Bridge first selected streamlined flat steel box girder, which increased the wind resistance and torsional strength of the bridge, and the steel consumption was small, and the maintenance was convenient and popularized. In 1970, the Danish Small Seaweed Bridge used the air drying device inside the box for the first time, which played a good anti-seismic effect on the steel box girder. In 1995, the Kobe Earthquake in Japan and the successful seismic design of the Akashi Kaikyo Bridge that had withstood the earthquake test made the technology of the suspension bridge develop unprecedentedly in all aspects.

Development: With the rapid development of the world economy, especially from the 1980s to the end of the 20th century, the construction of suspension bridges all over the world were to the heyday, and 17 suspension bridges with a span of more than 1000m were built. The famous suspension bridges built in the world include: the Mackenac Bridge and the Velazano Bridge built in the United States around the 1960s; the Hengbo Bridge was built in the UK in the 1980s; the Sea Bridge in Denmark in the 1990s. Japan built the Nanbeizan Seto Bridge, which propelled the suspension bridge span from 1000m in the 1930s to nearly 2000m, which is another major breakthrough.

6. Combined System bridge

In addition to the basic forms of the above bridges, several forms of combined structures,

FIG.4-48 Golden Gate Bridge

FIG.4-49 Akashi Kaikyō Bridge

such as a combination of beams and arches, a cable-stayed cable and a suspension cable combination system, are also used in engineering practice.

The strength characteristics of various forms of bridges should be fully utilized to give full play to their superiority.

4.6.4 Piers and Abutments

The supporting structure of the bridge is the abutment and the pier.

The abutment is the supporting structure of the bridge ends at both ends of the bridge, and is the connection point between the road and the bridge.

The pier is the intermediate support structure of the multi-span bridge.

1. Types of Piers

Function: The role of the pier is to support the vertical and horizontal forces transmitted from the left and right upper structures through the support.

Because the pier is built in the river, it has to withstand the pressure of the water, the wind above the water surface and the possible ice pressure, the impact of the ship. Therefore, the pier must have sufficient strength and stability in the structure. The interaction between the pier and the river should be considered in the arrangement, that is, the problem of river washes the pier and backwater. The requirements for navigation and traffic should be met in space.

Types: The types of piers commonly used in highway bridges can be divided into solid (gravity) piers, hollow piers and pile piers according to their structural forms, as shown in FIG.4-50.

- **Solid Pier**

Feature: It relies on its own weight to bal-

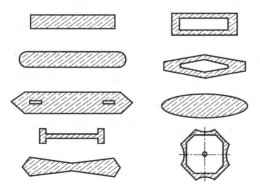

FIG.4-50 Different Cross Sections of Piers

ance external forces and maintain stability. It is generally suitable for bridges with large loads, or rivers with more ice and floating objects.

Disadvantage: The biggest disadvantage of this kind of pier is that the construction volume is large, so its dead weight and water-blocking area is also large. Sometimes in order to reduce the volume of the pier, the top portion of the pier is made into a cantilever type.

- **Hollow Pier**

Advantages: The hollow pier overcomes the shortcomings of the solid pier in many cases, and the concrete or reinforced concrete pier is made into a hollow thin-wall structure, which can save construction materials and reduce weight.

Disadvantages: The disadvantage is that it cannot withstand the impact of floating objects.

- **Pile Pier**

Due to the extensive use of large-aperture bored pile foundations, pile-type piers have been widely used in bridge engineering.

This structure extends the pile foundation all the way up to the bridge span structure. The pile top is poured into the pier cap, and the pile is used as a part of the pier body. The pile and the pier cap are made of reinforced concrete.

This structure is generally used when the bridge span is no more than 30m and the pier body is not higher than 10m. If the platform is built on the top of the pile and the pile is built on the pile as the pier, it will become a column pier.

Column piers can be single column or double column or multiple columns, depending on structural requirements.

2. Type of Abutment

The abutment is a supporting structure for the bridge ends at both ends, and can also connect roads and bridges.

Form of force: It has to withstand the vertical and horizontal forces transmitted from the support, and also retains the earth and revets the shore, and withstands the lateral earth pressure generated by the backfill and the load on the fill. Therefore, the abutment must have sufficient strength and avoid excessive horizontal displacement, rotation and settlement under load, which is especially important in statically indeterminate structure bridges.

- **Solid Abutment**

The U-shaped abutment is the most commonly used form of abutment, which is named after the support bridge spans the structure of the platform and the wing walls on both sides form a U-shape on the plane.

Feature: It is generally constructed with construction materials and has a simple structure.

It is suitable for bridges with a height of 8~10m or less and a slightly larger span.

Disadvantages: The disadvantage is that the abutment has a large volume and self-weight, which also increases the requirements for the foundation.

- **Embedded Abutment**

Feature: The embedded abutment is to embed the majority of the abutment body into the conical slope, and only leaks the toe cap to arrange the support and the upper structure. In this way, the abutment volume can be greatly reduced. However, because the slope protection in front of the abutment is a permanent surface protection facility, there is a possibility that the platform is exposed by the flood, so it is generally used for the shoal of the bridgehead.

4.6.5 Development Direction of Bridge Technology

- **Long-span bridges develop in a longer, larger, and softer direction**

The safety and stability of long-span bridges under the action of aerodynamics, earthquake and driving dynamics are studied. The cross-sections are made into various streamlined stiffening beams that meet the requirements of aerodynamics, and the stiffness of super-long span bridges is increased.

- **Development and application of new materials**

The new materials should be characterized by high strength and light weight. The research on ultra-high strength silicon salt and polymer concrete, high-strength two-way steel fiber reinforced concrete, fiber plastic and other materials replaces the steel and concrete used in bridges.

- **Development of the design stage**

In the design stage, the computer aided method of high-speed development, numerical simulation technology, is used to carry out effective rapid optimization and simulation analysis, and the intelligent manufacturing system is used to produce parts in the factory, and the bridge construction is controlled by GPS and remote control technology.

- **Large deep water foundation project**

At present, the world's bridge foundation has not exceeded 100m deep-sea infrastructure projects, and the next step is to practice the 100~300m deep-sea foundation.

- **Automatic monitoring and management**

After the bridge is completed and delivered, the bridge will be safely and normally operated through an automatic detection and management system. Once a fault or damage occurs, the damage location and maintenance measures will be automatically reported.

- **Pay attention to bridge aesthetics and environmental protection, pay attention to function development**

The bridge structure of the 21st century will definitely pay more attention to architectural art modeling, attach importance to bridge aesthetics and landscape design, attach importance to environmental protection, and achieve the perfect combination of human landscape and environmental landscape.

4.7　Development and Utilization of Subsurface

The 19th century is the century of the bridge, and the 20th century is the century of high-rise buildings. Scientists predict that the 21st century will be a new century in which humans rationally develop and utilize underground space.

In today's world, human beings are developing underground, ocean, and the universe. The underground development can be summarized as: underground resource development, underground energy development, and underground space development. The use of underground space is also constantly evolving.

In the 1980s, the International Tunnellling Association (ITA) proposed the slogan of "Developing underground space vigorously and starting a new era of human burrowing." In line with the trend of the times, many countries regard underground development as a national policy. For example, Japan proposes to expand underground and enrich the country by ten times. The history of the use of underground space echoes the history of human civilization, and it can be divided into four eras:

The first era: from the emergence of mankind to the ancient times of 3000 BC. Natural caves have become the place for humans to prevent cold and heat, shelter from the wind and hide wild animals.

The second era: the ancient period from 3000 BC to 5th century. The Egyptian Pyramids and the ancient Babylonian water tunnels are examples of architecture in this era. The tombs and underground granaries in the Qin and Han Dynasties in China have a certain level of technology and scale.

The third era: the medieval era from the 5th to the 14th century. The world's ore mining technology has emerged and promoted the development of underground engineering.

The fourth era: from the 15th century to modern times. In the industrial revolution in Europe and America, Nobel invented yellow explosives (TNT) and became a powerful weapon for the development of underground space. In the Meiji era of Japan, tunnel and railway technology began to be introduced and developed. In the Nordic countries, such as Sweden, in terms of underground space utilization, in addition to the basement of the house and urban facilities, many urban structures using solid rock caves can be seen, including underground streets, subway tunnels, public facilities ditch, parking lots, air-conditioning facilities and underground sewage treatment plants, as well as a series of underground shopping malls.

Japan's land is narrow, although the comprehensive utilization of underground space is later than that of countries such as Northern Europe, the scale of construction of underground streets, underground stations, underground railways, and underground shopping malls can be considered as the world's leading position.

4.7.1　Underground industrial buildings

Underground power stations, nuclear power plants, underground power plants, underground garbage incineration plants, etc. are all underground industrial buildings.

1. Underground power station

Definition: A underground power station is a hydropower station with a plant underground.

Advantages: Underground power stations can make full use of the terrain, especially in narrow valleys, and it is very cost-effective to build stations and arrange generator sets underground. The power station is built underground and can get more head pressure, and it can generate electricity when the water level is low during the dry season. The pressure tunnel of a general hydropower station is built in hard and complete rock to simplify the lining structure.

Types: Underground hydropower stations include a series of buildings and structures on the ground and underground, which can be summarized as two parts: dam and power station. The dam belongs to a large-scale hydraulic construction. The power station mainly includes the main plant, the auxiliary plant, the power distribution room and the switch station.

FIG.4-51　Three Gorges Underground Power Station

2. Underground nuclear power plant

Definition: Systems and equipment that convert nuclear energy released by nuclear fission into electrical energy, often referred to as nuclear power plants.

Disadvantages: There are a large amount of radioactive materials in the reactor of a nuclear power plant. If it is released into the external environment during an accident, it will cause great harm to the ecology and the people. Therefore, the nuclear power plant is generally built underground.

present situation: A nuclear power plant is a high-energy, low-consumption power station. By the end of 1986, 397 nuclear power plants had been built in 28 countries and regions around the world. According to the statistics of the International Atomic Energy Agency, it is estimated that 58 countries and regions will build nuclear power plants by the beginning of the 21st century. The total number of power stations will reach 1000, the installed capacity will reach 800 million kilowatts, and the nuclear power generation will account for 35% of the total power generation.

3. Underground factory

It is not only a strategic need but also a trend of economic development to hollow out the mountain or dig it down on the ground to put some factories underground.

The underground factory has the following **advantages**:

a. Shorten the production process

For example, in the underground printing factory of the newspaper, in order to shorten the time from the editing to the release of the newspaper, it is necessary to set the printing factory beside the editorial department, but the editorial department is often in the center of the city. It is a good choice to place the printing factory in the underground of the editorial department to reduce costs.

FIG.4-52   Underground plant invested by Dalian Guangyang Technology Group

b. Reduce noise

As people's environmental requirements increase, noise control is a big issue we face. If some of the more noisy production facilities are moved to underground, the formation can cut noise and greatly purify the environment.

c. Constant temperature and constant humidity

If the depth of the embedding is 3~5 m, the temperature in the formation will hardly change all the year round, and the humidity will be almost constant. It is highly advantageous to bury a production facility that requires these features, such as a freezing plant or a refrigerated warehouse.

d. Avoid destroying the landscape

A large number of ground buildings have destroyed the green landscape, and some produc-

tion facilities have been moved to the ground as much as possible, which can greatly reduce the damage to greening.

4.7.2　Underground storage building

1. Underground warehouse

Advantages: Since the underground environment has advantages for the storage of many substances, the thermal stability, airtightness of the underground environment and the good protective performance of underground buildings provide very favorable conditions for the construction of various warehouses underground.

Due to population growth and concentration, countries around the world are facing energy, food, water supply and radioactivity and other waste disposal issues.

present situation: Various types of underground storage facilities have occupied a large proportion in the total construction of underground works. In terms of energy storage and energy conservation in the development and utilization of underground space, Nordic, Scandinavia, the United States, the United Kingdom, France and Japan have achieved remarkable results.

2. Underground parking lot

With the development of the national economy, the number of vehicles owned by households is increasing. At the same time, people are increasingly demanding the living environment. Reducing the number of parking on the ground, building an underground parking plant, and better solving the diversion of people and vehicles are urgent problems to be solved.

About a quarter of the parking lots in Japan are underground parking lots.

4.7.3　Underground civil buildings

The development and utilization of urban underground space has become an important part of modern urban planning and construction. Starting from the construction of underground streets, underground shopping malls and underground garages, some large cities have gradually developed into multi-functional underground complexes that integrate underground commercial streets, underground parking lots, underground railways and underground pipeline facilities.

An underground connecting passage is established between the underground layers of various buildings, or a separate single room, forming an underground road that is generally narrow and adjacent with shops, offices, parking facilities, etc, collectively referred to as *underground streets*.

The underground street is the most developed in Japan with a small territory and a large population. Tokyo's Yaesu Underground Street (As showed in FIG.4-53) is one of the largest underground streets in Japan. It has a length of about 6km and an area of $6.8 \times 104m^2$. It has 141 stores and is connected to 51 buildings, with more than 3 million people working every day.

FIG.4-53　Tokyo's Yaesu Underground Street

4.7.4　Civil Air Defense Engineering

Definition: Civil air defense engineering refers to underground protective buildings constructed separately to ensure wartime personnel and material shelter, civil air defense command, medical rescue, and a basement built in combination with ground buildings that can be used for air defense in wartime.

FIG.4-54　Civil Air Defense Projects in Yichun South Cha District

Role: The civil air defense project is an important facility to guard against sudden attacks by the enemy, effectively cover personnel and materials, and preserve the potential of war. It is an engineering guarantee that adheres to urban fighting and long-term support for the anti-aggression war until victory.

4.7.5　Development and utilization of underground space

Rational development and utilization of underground space is an effective way to solve urban limited land resources and improve urban ecological environment.

The development and utilization of the underground space in Boston, USA is a good example:

Boston is an ancient city in the history of the United States and the busiest city. Its urban traffic problems have severely restricted the development of the city. The vitality of the urban center is reduced, the economic growth is slow, and the quality of the urban environment is declining. Built in 1959, Boston's Central Avenue is an elevated 6-lane road that runs directly through the city centre. It was designed to transport 75000 vehicles a day, and now it has reached

200000 units, making it the most crowded urban transportation line in the United States. Boston now has more than 10 hours of traffic congestion per day and the traffic accident rate is four times that of other cities. It is estimated that the annual loss due to poor traffic is about $500 million. The only viable option is to build an underground expressway below the existing central street. After 15 years of renovation of the Boston Central Avenue Tunnel, it was completed in 2004 and has a profound impact on urban construction.

At present, the development depth of urban underground space has reached about 30m, and the potential of underground space resources is very large. Therefore, the development of underground engineering can not only open up a broad space for human survival, but also have good benefits in social, economic, and environmental aspects, and is the only way for modern urban construction.

4.8   Water Resources and Hydropower Engineering

From ancient times to today, in order to meet the needs of survival and development, mankind has taken various measures to regulate the waters of nature to prevent floods and droughts, develop and utilize water resources.

Definition: The body of knowledge that studies the technical theories and methods of such activities and their objects is called hydraulic science. The project used to regulate the surface water and groundwater in nature to be built for rational use is called hydraulic engineering.

Feature: the hydraulic structure is affected by water, the working conditions are complex, and the construction is difficult.

The natural conditions of hydrology, meteorology, topography and geology vary from place to place, and the hydrological and meteorological conditions are contingent. Therefore, the design of large-scale water conservancy projects always has its own characteristics and is difficult to be uniform.

Large-scale water conservancy projects have large investment and long construction period, which have a great impact on society, economy and environment.

4.8.1   Farmland Water Conservancy Project

Farmland water conservancy projects mainly include **irrigation projects** and **drainage projects** to solve irrigation, drainage and soil improvement work in farmland.

1.   Irrigation and drainage

When farmland moisture does not meet crop needs, water should be added, which is irrigation. When there is too much water, water should be reduced, called drainage. Irrigation is to increase water when farmland moisture does not meet crop needs. Drainage is when water is too much, then water should be reduced.

- **Irrigation method**

a. Ground irrigation

Direct drainage of surface water into farmland is called **natural irrigation**. Due to the topography, it is sometimes necessary to raise the stage of the water source to divert water into the field. With the increasing contradiction between water shortages and increasing water demand, flood irrigation is gradually being replaced by modern water-saving irrigation models that focus on precise irrigation.

b. Sprinkler irrigation

Types: Sprinkler irrigation currently uses three types of fixed, semi-fixed and mobile, mainly consisting of pipes and nozzles.

When irrigation is required, the valve on the pipe is opened and the pressurized water is sprinkled from the nozzle to form a uniform water droplet that is scattered in the field.

Advantages: Saving water, and in addition to a small amount of evaporation, most of the water is effectively used for crop growth. It will neither raise the water table, nor make the land saline. No drainage is required during the irrigation period.

Disadvantages: The initial basic investment is large and the management is more complicated.

c. Drip irrigation

Drip irrigation refers to the construction of a special pipeline network in the ground. A drop of water is evenly dropped into the soil near the root zone of the crop through an drip emitter.

FIG.4-55 Ground Irrigation

FIG.4-56 Sprinkler Irrigation

FIG.4-57   Drip Irrigation

- **Irrigation water source**

Irrigation water sources mainly include natural river water, reservoirs, lakes, ponds, purified sewage, melted snow water, and groundwater.

- **Farmland drainage**

The main purpose: The main purpose of farmland drainage is to remove groundwater and reduce groundwater levels.

Therefore, it is required that the gutters must be dug to a certain depth and have appropriate longitudinal slopes to discharge water into rivers, lakes or oceans. If the outlet of the drainage system is higher than the water level of the river, lake and sea, it can be drained by itself and the management is more convenient. Otherwise you need to use a water pump.

Types: Drainage systems can be divided into open ditch drainage systems and underdrain drainage systems. The dark ditch can not only discharge surface water but also drain groundwater, but the cost is too high, and it can only be used in special circumstances. Generally, large-area farmland drainage uses open ditch drainage.

2. Water intake project

Function: The function of the water intake project is to introduce river water into the channel to meet the needs of farmland irrigation, hydropower generation, industrial and domestic water supply.

- **Free-dam water intake**

Types: The dam-free water intake channel is generally composed of three parts: the intake sluice, the sluice gate and the diversion dyke. In order to facilitate water diversion and prevent sediment from entering the channel, the intake sluice should generally be located in the recess of the river channel. In general, the design draw water flow does not exceed 30% of the river flow.

Although the dam-free water intake is simple, it does not have the ability to regulate the water level and flow of the river. It completely depends on the water head difference between the river water level and the water channel. Therefore, the water diversion flow is greatly affected by the change of river water level.

- **Dam water intake**

The construction of a dam or a backwater dam is a water intake method that can regulate the water level of the river and cannot regulate the large flow. When the river is rich in water resources but the water level is low, a backwater building can be built on the river, the water level can be raised, and the water can be irrigated by self-flowing water to form a way of dam water diversion. The elevation of the dam or river gate is determined by the water level required for drainage in the irrigation area.

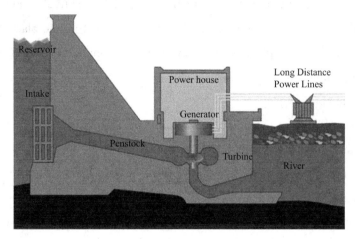

FIG.4-58   The Structure of Dam Water Intake

Advantages: The main advantage of dam diversion is that it can avoid the influence of river water level changes and stabilize the diversion flow. The main disadvantage is that the cost of building a dam is quite large, and the riverbed also needs suitable geological conditions.

- **Pumping water**

When the river is rich in water, but the irrigation area is high, and it is difficult or uneconomical to construct other self-flow diversion works, pumping water can be taken. In this way, the amount of main canal works is small, but the mechanical and electrical equipment and annual management costs are increased. The water diversion flow is determined by the capacity of the electrical equipment.

- **Reservoir water intake**

When the flow and water level of the river cannot meet the irrigation requirements, it is necessary to construct a reservoir at the appropriate location of the river to solve the problem of water supply.

In the case of water intake from reservoirs, buildings such as dams, spillways and intake gates must be built. The project is large and there is corresponding flood loss in the reservoir area, but the reservoir can make full use of river water resources.

3. Irrigation pump station and drainage pump station

- **Irrigation pump station**

When the location of the irrigation area is relatively high and the water of the water source cannot be self-flowed, the pumping station is generally set up on the upstream or downstream of the reservoir, or on the riverside or on the channel according to the local conditions.

a. Pump station planning and irrigation district division

Generally, when the irrigation area is small, the terrain is relatively simple, and the head is not large, a single-stage water pump is used to build a station in one place. If the irrigation area is large, the terrain is complex, and the elevation is not single, the district construction and multi-level water pumping are often used.

b. Pump station building layout

Types: The buildings of the pumping station generally include diversion canals, intake sumps, pump rooms, outlet pipes and outlet cisterns.

When the terrain between the water source and the outlet cistern is gentle, it is usually in the form of a diversion canal.

When the pumping station is located in a steep terrain or the irrigation area is close to the water source, the waterless channel arrangement is often adopted. The pumping station is built on the shore of the water source and directly draws water from the water source.

Feature: the pump room is greatly affected by the water level of the water source, and the flood problem is difficult to solve.

- **Drainage pump station**

a. Planning principles

Drainage must be carried out in zones, depending on local conditions.

Principles: high and low water separation, mainly self-discharge, supplemented by machine, drainage time is determined by the depth and time of flooding of various crops, and strives to drain in time.

b. Site layout

When drowning water can be collected in the storage area through the drainage main channel, it is advisable to build a larger pumping station. In order to make the unit of the drainage station small in installed capacity, low in cost, short in power transmission line, and convenient for centralized management.

The water network in the drainage area is dense, the drainage outlet is scattered, the terrain is uneven, the highland needs irrigation, and the low ground needs drainage, which need to build stations discretely. The decentralized construction of the station makes the pump station and the drainage channel have a small amount of work, and the effect is fast and easy to manage. This type of pumping station is preferably a combination of irrigation and drainage.

c. Arrangement of drainage pump station

There is no big difference between the building of the drainage pumping station and the irrigation pumping station. For the river, it is only the direction of the inlet and outlet. If the terrain is moderate, and the irrigation and drainage are both in the nearby farmland. After demonstration, the irrigation and drainage can be combined with the pumping station.

4.8.2 Hydropower Engineering

Hydroelectric power is a bit more prominent in water, it can be recycled, and it will not pollute the environment. The cost is much lower than the cost of thermal power generation. All countries in the world try to develop their own water resources.

1. Hydropower development methods and main types

In addition to the need for flow, hydropower requires a concentration drop. The natural concentration drop is only found in a few places, and that is the waterfall.

It is usually necessary to use manual methods to concentrate the drop, which can be divided into the following **ways:** dam-type hydropower station, diversion type hydropower station, pumped storage power station, tidal power station. Among them, dam-type hydropower stations and diversion hydropower stations are the most basic development methods.

● **Dam-type hydropower station**

Types: According to the relative position of the hydropower station of the dam-type hydropower station and the barrage or overflow dam, it can be divided into four types: riverbed, post-dam, overflow or hybrid.

a. Riverbed hydropower station

Feature: The characteristic is that only low dams are built, and the reservoir capacity and regulation capacity are small, mainly relying on the natural flow of rivers to generate electricity. Due to the large amount of abandoned water, the utilization of water energy is greatly limited, and the comprehensive benefits are small, but the difficulty of flooding loss and resettlement is also small.

This type of hydropower station is suitable for construction in plains or hilly areas, where the slope of the river is slow and the elevation of the water level will significantly increase the loss of urban and rural flooding on both sides of the river (FIG.4-59).

b. Post-dam hydropower station

Feature: This form is characterized by the fact that the power plant of the hydroelectric

FIG.4-59   Riverbed Hydropower Station

power station is close to the downstream of the water retaining dam, and the power generating water pressure pipe directly enters the turbine room in the hydropower plant through the dam.

Therefore, the structure of the plant is not limited by the water head, and the head depends on the height of the dam (FIG.4-60).

c. Overflow hydropower station

When the hydropower resources are located in the alpine valley area, the plant and the spillway often cannot be arranged at the same time. This form combines the overflow dam with the plant, overflows or drains from the plant, and then further develops to not directly overflow the hydropower station, but directly flows the flood to the downstream riverbed (FIG.4-61).

FIG.4-60   Post-dam Hydropower Station   FIG.4-61   Overflow Hydropower Station

- **Diversion power station**

The diversion power station mainly uses water diversion channels to concentrate the water heads. However, strictly speaking, from the way of concentrating the water head, most of the water heads are concentrated by the dam body, and some of the water heads are concentrated by the water diversion channel.

- **Pumped storage power station**

Feature: Pumped-storage power stations are special-purpose hydropower stations that redistribute energy over time.

In the latter half of the night, when the power system load is at a low point, the hydropower station, especially the excess power of the atomic power station, uses the pumped storage method to store energy in the reservoir, and the unit draws water from the downstream into the reservoir. At the peak of the power system, the amount of water stored is used to generate electricity, which converts water into electricity. But there will be energy loss in the middle.

- **Tidal Hydropower Station**

Tidal power generation uses the potential energy of the water level difference caused by rising and falling tides to generate electricity. According to calculations, the world's ocean tidal energy reserves are 2.7 billion kilowatts. If all are converted into electrical energy, the annual power generation will be about 1.2 trillion $kW \cdot h$. Compared to other renewable energy sources.

The outstanding advantage of tidal energy is its ability to generate uninterrupted power.

2. Hydropower station buildings

• Arrangement of hydropower station buildings

a. Arrangement of buildings of riverbed hydropower stations

This arrangement is suitable for lower heads, generally below 30~40m, and is built on a relatively gentle section of the middle and lower riverbed of the river or irrigation channel.

Feature: It is characterized by the fact that the plant and the dam are built side by side in the riverbed, and the plant becomes part of the water retaining structure.

b. Arrangement of buildings of post-dam hydropower station

When the water head is high, generally more than 30~40m, due to the pressure, the weight of the plant itself is not enough to maintain its stability. Therefore, the plant is located downstream of the barrage and is not subject to upstream water pressure.

c. Arrangement of buildings of drainage hydropower stations

Due to the topography and geological conditions, the arrangement of the power station or the dam-free water cannot be arranged after the dam. The water flows through the inlet into the pressurized tunnel, through the surge tank to the pressure water pipe, and then to the turbine in the plant.

• The role of hydropower station buildings

a. Water retaining building

Generally, it is a dam or gate to cut off the river and concentrate the drop to form a reservoir.

b. Water discharge building

Used to vent excess flood or release water to reduce reservoir water level, such as spillway, flood tunnel, or drain hole.

c. Intake buildings of hydropower stations

It is the introduction of water into the inlet of the water channel

d. Hydropower station water diversion building

It is used to introduce water from the reservoir into the turbine. According to the topography, geology, hydrometeorology and other conditions of the hydropower station and the type of hydropower station, open channels, tunnels and pipelines can be used. Sometimes the water diversion channel also includes grit chambers, aqueducts, culverts, inverted siphons, bridges and buildings that direct water flow from the waterwheel machinery to the downstream.

e. Power generation, substation and distribution buildings

The structure includes a plant for installing a hydroelectric generator and its control equipment, a transformer field for installing a transformer, and a switch station for installing a high voltage switch. They are grouped together as a plant hub.

4.8.3　Flood Control Project

Flood protection includes countermeasures, measures and methods to prevent floods from harming humans.

Subjects of Major Studies: It is a branch of water science, and its main research objects include the natural laws of floods, the conditions of rivers and floodplains, and their evolution. The basic content of flood control work can be divided into construction, management, flood control and scientific research.

Flood is a natural phenomenon that often causes flooding of river valleys, alluvial plains, estuary deltas and coastal areas along rivers. Due to the periodic and random nature of flood phenomena, as well as changes in the natural environment and human activities, the extent and timing of flooding of these zones are both regular and unfixed and contingent. However, most of these areas threatened by flooding are still exploited by humans, and flood control problems have emerged.

1. Flood Control Planning

Flood control planning refers to the overall deployment to prevent floods in a certain river section or area. It includes river basin flood control planning for rivers and lakes determined by the state. Flood control planning is the basis for river and lake management and flood control engineering facilities construction.

The objectives, guidelines, policies and the importance of the protection zone of the national economic construction are the basis for the preparation of the plan.

Principles: Overall planning, comprehensive utilization, storage and venting, different treatment super-standard floods, and flood control projects combined with non-engineering measures for flood control.

Contents: collecting data, flood analysis, formulating planning objectives and flood control standards, formulating flood prevention plans, determining the height of flood control, and analyzing flood control benefits.

2. Flood Control Project

Flood control areas refer to areas that may be flooded, and are divided into floodplain areas, flood storage and flood protection areas.

Floodplain refers to the area where flooding is not covered by engineering facilities.

The flood storage and detention area refers to low-lying areas and lakes that temporarily store floods outside the piggyback of riverbank.

Flood protection zone refers to the area protected by flood control engineering facilities within the flood control standards.

Definition: The flood control project is a project constructed to control and defend against floods to reduce flood damage. It is mainly composed of dykes, river improvement projects, flood diversion projects and reservoirs.

Comprise: Flood control tasks in a river or an area are usually undertaken by an engineering system consisting of a combination of measures. Generally, reservoirs are built in the valleys of the upper and middle reaches of the tributaries to intercept floods and regulate runoff. The hilly area has extensively carried out soil and water conservation, water storage and soil conservation, and improved ecological environment. In the middle and lower reaches of the plain, build dykes, rectify rivers and control the estuary.

4.8.4 Development trends and prospects

1. Development trend

In most countries in the world, there are populations that are growing too fast, and the use of insufficient water resources, urban water supply shortages, energy shortages, and deterioration of the ecological environment are closely related to water. Flood prevention and control and utilization of water resources have become major issues in contemporary social and economic development.

The development trend of water conservancy projects is mainly:

a. The engineering measures for flood prevention are further integrated with non-engineering measures, and non-engineering measures are increasingly important.

b. The development and utilization of water resources will be further developed into a comprehensive and multi-objective development.

c. The role of water conservancy projects is not only to meet the needs of the growing people's lives and industrial and agricultural production, but also to protect and improve the environment.

d. Large-area, large-scale water resources allocation projects will be further developed.

e. Due to new exploration technologies, new analytical calculations and monitoring and testing methods, as well as the development of new materials and processes, complex foundations and high-head projects will be developed, and local materials will be more widely used, the

cost of the goods will be further reduced.

f. The unified management and unified dispatch of water resources and water conservancy projects will be gradually strengthened.

2. Prospects

Water is the source of human life. However, it is reported by 2025, the gradual lack of water resources will threaten the food supply of nearly 3 billion people worldwide and the health and productivity of swamps and other ecosystems around the world. About 450 million people in 29 countries and regions, including China, have been facing water shortages.

To address the dilemma between agricultural production and environmental protection, the world's most influential natural conservation, irrigation and food security organization has established an international science and policy coalition – the Water, Food and Environment Dialogue.

Many countries in the world have large-scale water transfer projects. For example, Israel's 11-year North-South Water Transfer Project has alleviated the harsh ecological and environmental conditions that constrain the development of the southern region, turning the vast desert into an oasis, thus driving the economic and social development in the south. However, the water transfer project is a huge and complex project that takes into account various factors. A slight inadvertent will cause great losses. India hopes to transform the Ganges River and connect rivers across the country with artificial canals, eventually achieving the goal of transporting water from the north to the southeast. The Ganges River mainly flows through India and Bangladesh. Once the Ganges River water is completely diverted, Bangladesh will become a dry place in the dry season, and there will be insufficient domestic water. In addition, the plants in the Ganges Triangle will be greatly damaged. Disasters are inevitable.

Question

- What issues to consider during the construction of the road ?
- With the development of high-speed railways, will the maglev railway disappear? Why?
- Please consider the advantages and disadvantages of high-speed railway construction.
- What are the issues that need attention when lighting road tunnels?
- What methods are included in subway tunnel construction generally?
- What structural forms are included in the bridge?
- What are the design priorities of the bridge project?
- What are the advantages of developing and utilizing underground space?
- What are the methods of irrigating farmland? What are the characteristics?

- Take the Three Gorges Hydropower Station as an example, please consider the advantages and disadvantages of hydropower station construction.

Reference List

[1] Liu Gang, Foundation Engineering [M]. Beijing: China Building Materials Industry Press, 2000.

[2] Yang Chunfeng, Road Engineering [M]. Beijing: China Building Materials Industry Press, 2000.

[3] Qian Binghua et al., Airport planning and design [M]. Beijing: China Building Industry Press, 2000.

[4] Yan Xikang, Civil Engineering Construction [M]. Beijing: China Building Materials Industry Press, 2000.

[5] Li Zuomin, Traffic Engineering [M]. 2nd edn. Beijing: People's Communications Press, 2000.

[6] Zhang Fengxiang, et al., Caissons [M]. Beijing: China Railway Press, 2002.

[7] Gao Mingyuan, Yue Xiuping. Building Water Supply and Drainage Engineering [M]. Beijing: China Building Industry Press.

[8] John S Scott. Dictionary of Civil Engineering[M]. 4th edn. London: Penguin Books Publisher, 1991.

[9] Ye Zhiming, Introduction to Civil Engineering[M]. 3rd edn. Beijing: Higher Education Press, 2009.

CHAPTER 5
CONSTRUCTION OF CIVIL ENGINEERING

5.1 Earthwork Construction

Earthwork is an important construction process for foundation construction, which is characterized by *large engineering volume* and *complicated construction conditions*.

Therefore, before the construction of earthwork, a reasonable construction plan should be determined according to the engineering and geological conditions. The smooth progress of earthworks can not only improve the labor productivity of earthwork construction, but also create favorable conditions for the construction of other projects.

FIG.5-1   Earthwork construction

5.1.1 Classification and properties of soil

- **Classification methods:** depositional age, particle gradation, compactness, and liquidity index.

In the civil engineering construction, the soil is divided into ***eight categories*** according to the difficulty of excavation of the soil: soft soil, ordinary soil, hard soil, gravel soil, soft rock, secondly hard rock, solid rock, special solid rock.

- ***Properties***: weight, natural water content, compactness, permeability coefficient, shear strength, looseness, compressibility.

5.1.2 Site grading

- ***Definition***: The leveling of the site is to transform the natural ground into the plane that people demand.
- ***Types***: First leveling the entire site, then excavating the foundation pit of the building;

First excavating the foundation pit of the building, then leveling the site;

While leveling the site, excavating the foundation pit.

Before the site is leveled, the design elevation of the site must be determined, the quantity of excavation and filling work should be calculated, the balance of the excavation and filling should be determined, the earthmoving machinery should be selected, and the construction plan should be formulated.

5.1.3 Mechanized construction of earthwork

Earthwork should be mechanized as far as possible to reduce heavy physical labor and improve the construction speed.

- ***Construction Machinery***: bulldozer, scraper and excavator.

The ***bulldozer*** is flexible and convenient in operation, small in working surface, fast in speed, easy to transfer, and wide in application.

FIG.5-2 Bulldozer

FIG.5-3 Scraper

FIG.5-4   Excavator

The **scraper** is easy to operate, not restricted by terrain, can work independently, travel fast, high production efficiency.

The **excavator** is also called single bucket excavator, has strong excavating ability, high working efficiency, good versatility, but poor mobility.

● **_The basis for selecting earthmoving machinery_**: the type and scale of earthwork; geological, hydrological and climatic conditions; mechanical and equipment conditions; construction period.

5.1.4   Dewatering

● **_The necessity of dewatering_**

If the groundwater level is high and the bottom of the foundation pit is lower than the groundwater level, the soil-aquifer is often cut off during the excavation of the foundation pit or trench, and the ground water will continuously seep into the pit.

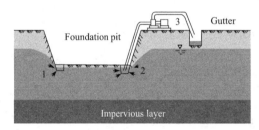

FIG.5-5   Collecting puddle precipitation

In order to ensure the normal construction, to prevent slope collapse and the decline of foundation bearing capacity, it's necessary to do a good job in foundation pit dewatering.

● **_Types_**: collecting puddle precipitation, light well point precipitation.

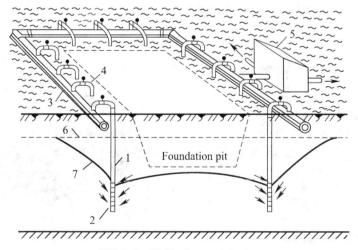

FIG.5-6   Well point precipitation

The collecting puddle precipitation method is generally suitable for the condition that the precipitation depth is small and there is no flowing sand in the stratum. If the precipitation depth is large, or there is flowing sand in the stratum, or in a soft soil area, the well point precipitation method should be adopted as far as possible. No matter which method is adopted, the precipitation work will continue until the foundation construction is completed and backfilled.

5.1.5   Earthwork filling and compaction

In order to ensure the quality of the filling project, the soil material and filling method must be selected reasonably according to the requirements of the filling. The filling material should meet the design requirements to ensure the strength and stability of the filling.

- *Requirements*: The selected filler should be soil and stone with high strength, low compressibility, good water stability and easy construction. Filling should strictly control the water content so that the water content of the soil material is close to the optimum water content of the soil.

- *Compaction Methods*: rolling, tamping and vibratory compaction.

Rolling is suitable for large-area filling works; *tamping* is mainly used for small-area filling, which can compact cohesive soil or non-cohesive soil; *vibratory compaction* is mainly used for compacting non-cohesive soil.

FIG.5-7   Rolling

The quality of fill compaction is related to many factors, among which the main influencing factors are compaction work, soil water content and thickness of each layer.

5.2   Foundation Construction

General industrial and civil buildings use natural shallow foundations as far as possible. When the soil layer is weak and the buildings have higher requirements for deformation and stability, a deep foundation is adopted.

The deep foundation has high bearing capacity and requires special methods for construction. The construction technology is complicated, the construction period is long, and the cost is high. **Common deep foundations**: pile foundation, pier foundation, sinking foundation, caisson foundation and underground continuous wall.

5.2.1　Pile foundation

- **Compose**: pile and bearing cap connecting pile top.
- **Features**: high bearing capacity, good stability, small and uniform settlement.
- **Types**: According to the manufacturing method of the pile, it can be divided into **precast pile** and **cast-in-place pile.**

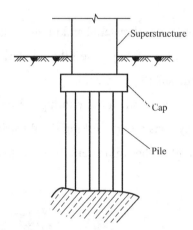

FIG.5-8　Pile foundation　　　　　　**FIG.5-9**　Precast pile

Precast reinforced concrete piles are durable and resistant to groundwater and humid environments and can withstand large loads. The construction methods of reinforced concrete precast piles include hammering method, vibration method, water rushing method and static pressure pile method.

Compared with precast piles, the cast-in-place piles have the **advantages** of no need to pick up piles and cut piles; economical, convenient construction and low noise. However, there are also **shortcomings** such as long technical intervals, no immediate load, and easy necking and broken piles in soft soil foundations. The construction methods of the cast-in-place pile include dry-working bored piles, mud retaining wall-forming bored piles, casing-forming bored piles and blasting hole-filled piles.

FIG.5-10　Cast-in-place pile

5.2.2　Pier foundation

- **Definition**: a foundation formed by pouring concrete in a large diameter hole which is manually or mechanically excavated in the ground.

- ***Diameter***: between 0.80 and 5.00 m, usually about 0.80~2.50 m.

- ***Component***: the pier cap, the pier body and the extension head.

- ***Advantages***: The pier foundation has a large diameter and a high bearing capacity;

Easy to guarantee the quality of pouring concrete, low cost and fast construction speed.

- ***Disadvantages***: The main problem of the pier foundation is the difficulty brought by the construction when the groundwater level is high and the collapse accident caused by the flow of sand. The large amount of earth unloading will also cause the looseness of the soil in the hole wall, thus reducing the structure of the soil strength.

FIG.5-11   Pier foundation

5.2.3   Sunk well foundation

- ***Definition***: Sunk well foundations are usually round or square tubular structures made of reinforced concrete.

- ***Component***: the blade foot, the wellbore, the inner partition wall, the back cover and the top cover.

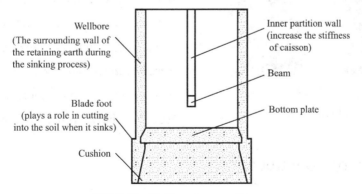

FIG.5-12   Sinking structure diagram

- ***Construction process***:

(1) The well body of the sunk well foundation is first placed on the ground during construction;

(2) Then digging the soil out of the well;

(3) As the soil in the well gradually deepens, the sinking sinks gradually;

(4) When sinking to a certain depth, connect the long shaft wall on the ground;

(5) Continue to dig and sink until it sinks to the design elevation;

(6) Finally, the back cover of the caisson or the filling of the top of the well is carried out.

5.2.4　Underground continuous wall

- **Construction process**: construction of the guide wall, division of the trench, excavation of the trench, the circulation process of the mud counterfort, joint construction of the trench, processing and hoisting of the steel cage, and pouring concrete.

- **The key to the construction**: the trench, the mud counterfort and the underwater pouring concrete.

FIG.5-13　Underground continuous wall
(Provided by China Geotechnical Network)

- **Basic principle**: To use a special trenching equipment on the ground to excavate a long and narrow deep trench along the periphery of the deep excavation project, and place a steel cage in the trench. Pouring the concrete and building a continuous underground wall arch to retain soil, load or intercept water and seepage.

- **Advantages**: the wall has high rigidity and can withstand large soil pressure; the vibration during construction is small, the noise is low; the anti-seepage performance is good; it is suitable for various geological conditions.

- **Disadvantages**: more equipment is needed, the construction process is more complicated, and it needs to have a certain level of technology.

5.3　Masonry Construction

Masonry engineering refers to the construction of ordinary clay bricks, silicate bricks, stones and various blocks. The materials used in masonry engineering are low in price, convenient in material extraction, and good in thermal insulation and durability, but the construction is high labor intensity and low productivity.

The use of small and medium-sized blocks and lightweight partition walls is the development trend of masonry engineering. The continuous development of new walls promotes the progress of traditional masonry construction technology.

FIG.5-14   Masonry construction

5.3.1   Masonry materials

The materials used in masonry works are mainly ***brick***, ***stone*** or ***block*** and ***masonry mortar***.

- ***Brick***: Ordinary clay bricks, autoclaved lime-sand bricks, fly ash bricks and load-bearing clay hollow bricks.
- ***Size***: The size of common standard clay brick is 240 mm × 115 mm × 53 mm.

FIG.5-15   Ordinary clay bricks

FIG.5-16   Clay hollow bricks

5.3.2   Masonry construction technology

> **Bricklaying construction**

- ***Construction process***: copy the flat line, laying bricks, vertical skins, wall hanging lines, ash bricks, joints, cleaning walls.

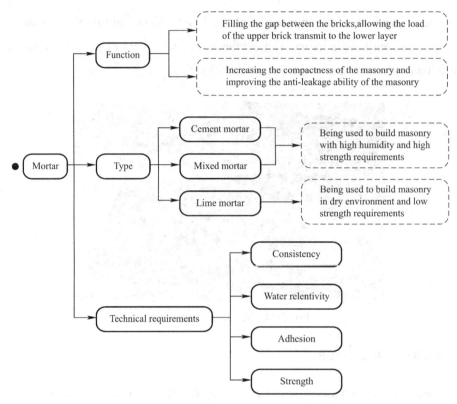

FIG.5-17   Mortar

FIG.5-18   Laying Bricks

FIG.5-19   Ash Bricks

- *Requirements*: Horizontal and vertical, full mortar, uniform gray joints, up and down staggered seams, internal and external framing, and solid joints.

Up and down staggered seams means that the vertical joints of the upper and lower bricks of the brickwork should be staggered to avoid the upper and lower joints. The *joint* refers to the temporary discontinuity that can be set up by the adjacent masonry at the same time, which is convenient for the joint between the first masonry and the rear masonry.

- *Masonry form*: flemish bond, flemish garden wall bond, plum blossoming method.

FIG.5-20 Up and down staggered seams

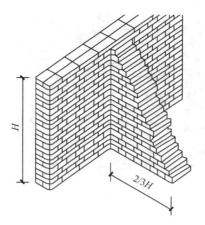

FIG.5-21 The joint

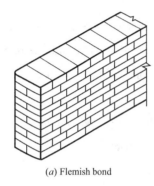

(*a*) Flemish bond

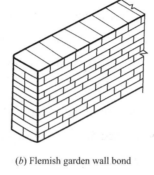

(*b*) Flemish garden wall bond

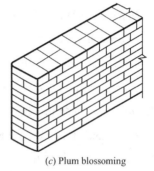

(*c*) Plum blossoming

FIG.5-22 Masonry form

> **Masonry construction**

Stone masonry includes both rubble and stone masonry. It is widely used in building foundations, retaining walls and bridge piers.

Rubble masonry should be divided into layers, and should be up and down staggered, built inside and outside, can not be used to build the middle of the side of the stone. When laying stone masonry, it should be placed smoothly. The thickness of the mortar should be slightly higher than the specified thickness of the mortar joint.

FIG.5-23 Masonry construction

> **Block construction**

● *Type*: Fly ash silicate blocks and normal concrete hollow blocks.

The block is large and heavy, and it should be hoisted by mechanical transportation. There-

FIG.5-24　Block Construction

fore, the block layout diagram should be drawn before lifting to guide the hoisting construction.

- ***Construction characteristic***: The number of blocks is large and the number of lifting times is correspondingly large, but the weight of the block is not large.
- ***Construction process***: Ashing, lifting block seating, correction, potting and bricking.

5.3.3　Winter construction of masonry

- ***Definition***: When the outdoor average temperature is expected to be lower than 5℃ for 5 consecutive days, or outside the winter construction period, the minimum temperature of the day is lower than −3 ℃ , the masonry construction should be built according to the requirements of winter construction.

The *materials* used in winter construction shall meet the following *requirements*:

(1) Bricks and stones should be cleaned of frost before laying bricks;

(2) The mortar should be mixed with ordinary Portland cement;

(3) Lime paste and clay puddle should be protected from freezing. If frozen, it should be used after melting;

(4) The sand used for mixing mortar shall not contain ice cubes and ice agglomerates larger than 1 cm in diameter;

(5) When mixing mortar, the temperature of water should not exceed 80 ℃ , and the temperature of sand should not exceed 40 ℃ .

Ordinary bricks should be properly watered and wet under normal temperature conditions. When laying at negative temperature conditions, if there is difficulty in watering, the consistency of the mortar should be increased appropriately.

In the winter construction, the surface of the masonry should be covered with insulation materials after daily masonry.

5.4　Reinforced Concrete Construction

Reinforced concrete engineering is composed of multiple types of work such as formwork, reinforcement and concrete. Due to too many construction processes, it is necessary to strengthen construction management, make overall arrangements, and organize reasonably to achieve quality assurance, speed up construction and reduce cost.

5.4.1 Formwork

- *Definition*: The formwork is a mold that allows the newly mixed concrete to meet the design requirements of the location and geometry to harden it into a reinforced concrete structure or component.
- *Components*: A formwork and a support.

The **formwork** is a model that forms the required form and size of the building; the **support** is a structure that supports and stabilizes the formwork and withstands the weight of the formwork, steel, concrete.

FIG.5-25   Formwork

- *Basic requirements*:

(1) Ensure that the shape and position of each part of the structure and components are accurate;

(2) Have sufficient bearing capacity, rigidity and stability;

(3) The structure is simple, and the assembly and disassembly is convenient;

(4) The seam of the template should not leak.

➢ **Type of Formwork**

In addition to the traditional wooden formwork, three series of industrialized formwork systems have been formed, namely, combined, tool and permanent.

The *combined formwork* refers to the template with strong applicability and versatility. It can be used to form the concrete structure. The *tool formwork* is a template for shaping the specific components (such as walls, column slabs, etc.) of the cast-in-place concrete structure. A *permanent formwork* is also known as a one-time template. After the template is poured into the concrete, the formwork is no longer removed, and the formwork is generally regarded as an integral part of the concrete member to share the load.

➢ **Formwork Design**

The **purpose** of the formwork design is to reasonably select the formwork material and the support system; to ensure that the formwork and support system have sufficient bearing capacity, rigidity and stability; easy to install and remove.

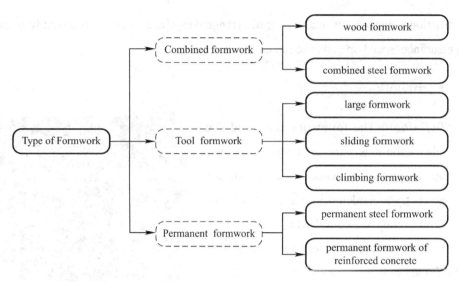

FIG.5-26 Template form

The **contents** of the formwork design include: selection, material selection, load calculation, structural calculation, structural design, drafting the installation and removal plan, and drawing the template.

  ➢ **Formwork construction**

(1) Installation of the formwork

During the installation process, the formwork and its support must be equipped with temporary anti-overturning measures. Installation of the upper formwork and its support should be noted that the lower slab should have the bearing capacity to withstand the upper load; the upper support should be aligned with the lower support and the slab; if the truss formwork method is adopted, the bearing capacity and rigidity of the supporting structure must meet the requirements during construction.

(2) Formwork Removal

The formwork of the cast-in-place structure and the concrete strength when the bracket is removed shall meet the design requirements; when there is no specific requirement for the design, the side formwork may be removed after the concrete strength can ensure that the surface and edges are not damaged by the removal of the formwork; The strength of the concrete required for removal should meet the requirements.

5.4.2 Reinforcement engineering

The steels commonly used in concrete structures and prestressed reinforced concrete structures are *steel*, *steel wire* and *steel strands*.

➢ **Banding of rebar**

● *Definition*: The steel bars that over-lap each other are tied together with thin iron wire.

● *Binding request*:

(1) When the steel bars are tied, the intersection of the reinforcement is fastened with iron wire;

(2) For the steel mesh between plate and wall, the intersections of the middle part can

FIG.5-27   Banding of steel bars

be crisscrossed to ensure that the position of the stressed steel bar does not shift;

(3) The stirrups of beams and columns shall be set perpendicular to the stressed reinforcement.

➢ **Welding of rebar**

● *Types*: Pressure welding and fusion welding.

Pressure welding includes flash butt welding, resistance spot welding and gas pressure welding; Fusion welding includes arc welding and electro-slag pressure welding.

(1) Flash butt welding is widely used for the connection of steel bars and the welding of prestressed steel bars and screw ends bars.

(2) Resistance spot welding is mainly used for cross-connection of small diameter steel bars.

(3) Arc welding is widely used in the welding of steel joints, steel frame welding, welding of fabricated structural joints, welding of steel bars and steel plates.

(4) Electro-slag pressure welding is often used in construction for the welding length of vertical or diagonal reinforcement in cast-in-place concrete structural members.

FIG.5-28   Welding of rebar

FIG.5-29   Mechanical connection of rebar

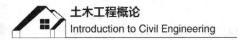
> ➤ **Mechanical connection of rebar**

- *Type of mechanical joints*: Extruded sleeve joints and threaded sleeve joints.

Reinforced mechanical joints shall meet the requirements for joint strength and deformation properties.

(1) The rebar extrusion joint

- *Definition*: The rebar extrusion joint is a joint formed by plastically deforming the steel sleeve by pressing force and closely engaging with the ribbed steel bar.

- *Applicability*: Being suitable for connection of large diameter deformed steel bars in vertical, horizontal and other directions.

- *Feature*: Compared with welding, the extrusion connection saves energy, is not affected by the weldability of the steel bar, is not affected by the weather, has no open flame, is easy to construct and has high joint reliability.

(2) The threaded sleeve joint

- *Definition*: The threaded sleeve joint is to machine the connecting end of the steel bar into a thread, and the two wire-reinforced steel bars are connected into a steel joint through a threaded connecting sleeve.

- *Feature*: It has a fast construction speed and stable quality and can be widely used in the connection of steel bars.

5.4.3 Concrete engineering

Concrete Engineering includes *construction processes* such as concrete *preparation*, *transportation*, *pouring*, *tamping* and *curing*. Concrete structures are generally the load-bearing parts of buildings, bridges, roads. Therefore, it is extremely important to ensure the quality of concrete works.

In recent years, *concrete admixtures* have developed rapidly, which has improved the performance and construction process of concrete. In addition, the development of new construction machinery and construction technology has also greatly improved the construction quality of concrete works.

> ➤ **Preparation of concrete**

- *Definition*: Concrete preparation refers to the mixing of various component materials that meet the technical requirements into a uniform concrete mixture with the required workability according to the specified mixing ratio.

In the preparation of concrete, mechanical mixing should be used except for manual mixing when the amount of work is small and the concrete is dispersed.

FIG.5-30 Self-propelled mixer   **FIG.5-31** Forced mixer

- *Determination of preparation strength*: The *concrete construction mixture ratio* should ensure the requirements of structural design on *concrete strength grade* and *concrete workability*, and should conform to the principle of rational use of materials and cement conservation. If necessary, it should also be resistant to *frost resistance* and *impermeability*.

- *Selection of concrete mixer*: When selecting a mixer, it depends on *the amount of work*, *the concrete slump*, *aggregate size*. It is necessary to meet technical requirements as well as economic benefits and energy conservation.

Concrete mixers are divided into two types according to their principle: *self-falling type* and *forced type*. The impact energy generated by the *self-propelled mixer* is small when it is stirred. The agitating effect of the *forced mixer* is stronger than that of the self-propelled mixer.

- *Determination of the mixing system*: In order to obtain a good quality concrete mix, in addition to the correct choice of mixer, the mixing system-*mixing time*, *mixing procedure* and *charging capacity* must be correctly determined.

➢ **Transportation of concrete**

- *Definition*: Transportation of concrete refers to the process of transporting concrete mixes from the preparation site to the pouring site.

- *Types*: Ground transportation, vertical transportation and surface transportation.

- *Basic requirements*: No segregation, guaranteed slump at the time of pouring, and sufficient time for pouring and tamping before the concrete is initially set.

- *Choice of transportation tools*: The choice of concrete transportation tools is mainly determined according to the characteristics of the structure, the amount of pouring required per unit time, the transportation distance, the transportation means, the running speed and capacity, and the site conditions.

FIG.5-32　Concrete mixing truck

FIG.5-33　Concrete pump

Concrete *ground transportation* generally adopts the concrete mixing truck when the ready-mixed concrete is used and transport distance is far.

Concrete *vertical transportation* generally adopts tower cranes, concrete pumps, fast lifting buckets and derricks.

Concrete *surface transportation*, small projects is mainly for two-wheeled trolleys; for large-area floor and other ground concrete projects, small motorized dump trucks are used.

> **Concrete pouring and tamping**

- *Purpose*: Ensure the homogeneity, compactness and integrity of the concrete structure.
- *Preparation work*: Before pouring, the formwork, falsework, rebar and embedded parts should be inspected and accepted.
- *Requirements*: Prevention of segregation, layered pouring tamping, correct placement of construction joints.
- *Concrete pouring method*:

(1) For the pouring of cast-in-place concrete frame structure, the construction layer and construction section should be divided first.

(2) The pouring of mass concrete structure can be divided into three types: overall layered pouring, section and layer pouring and slant layer pouring. The concrete pouring strength required by the overall layering method is large, and the concrete pouring strength of the slant layering method is small.

(3) When pouring concrete under water or mud, the tremie method is currently used.

- *Concrete compaction method*:

(1) Liquefaction by means of mechanical external force to overcome the internal force of the mixture;

(2) Appropriately adding water to the mixture to improve its fluidity, making it easy to

shape, and excess water and air will be discharged with separation method or vacuum operation method after forming;

(3) The high performance water-reducer will be added into the mixture to greatly increase the slump.

> **Concrete curing**

• **Definition**: The concrete is kept moist for a certain period of time under the condition that the average temperature is higher than 5℃ .

• **Types**: Artificial curing and natural curing.

Natural curing is divided into **two types**: sprinkler curing and spray film curing.

FIG.5-34   Concrete pouring

FIG.5-35   Concrete curing

Sprinkler curing refers to cover the concrete with straw curtains and often sprinkle water to keep it moist. The length of curing time depends on the type of cement, and the number of times of sprinkling is suitable to ensure the wet state. *Spray film curing* solution is suitable for high-rise building and large area concrete structures that are not easy to be sprinkled.

5.5   Prestressed Concrete Construction

The *working principle* of prestressed concrete is compressive pre-stress is applied to the concrete in the tension zone of the component before the component is subjected to the external load. When the tensile stress is generated by the external load of the component during the use phase, the compression pre-stress is firstly offset, which delays the occurrence of the concrete crack and also limits the crack initiation, thereby improving the crack resistance and rigidity of the component.

5.5.1   Overview

• **Feature**: Small cross section, high stiffness, good crack resistance and durability.

FIG.5-36   Prestressed concrete pile

- *The method of applying the pre-stressing stress*:

According to the order of comparison with the production of components, it is divided into two categories: *pre-tension method* and *post-tension method*. According to the tension method of the steel bar, it is divided into *mechanical tensioning* and *electric heating tensioning*.

To tension the steel bar in the tension zone, and the elastic contraction force of the prestressed steel bar is transmitted to the concrete component through prestressed steel bars or anchors, and prestress is generated.

- *Type of prestressed reinforcement*: Cold drawn low carbon steel wires, cold drawn steel bars, stranded wire, heat treated steel bars, and finished deformed bar.

- *Requirements of concrete*: The strength grade is not lower than C30.

5.5.2   Pre-tensioning construction

Construction process

(1) Before pouring the concrete member, tensioning the prestressing reinforcement and temporarily anchoring them to the stretching bed or the steel formwork.

(2) Pouring concrete members.

(3) Relaxing stress when the concrete reaches a certain strength, and the compression prestress is generated on the concrete by means of the bond between the concrete and the prestressing reinforcement.

> **Pre-tensioning construction equipment**

- *Stretching bed*: When the prestressed concrete members are produced by the stretching bed, the prestressing reinforcement are anchored on the beams of stretching bed, and the stretching

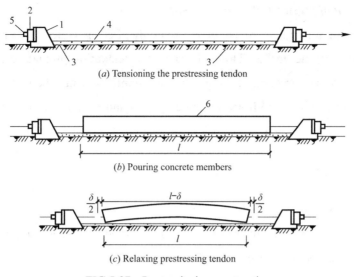

(a) Tensioning the prestressing tendon

(b) Pouring concrete members

(c) Relaxing prestressing tendon

FIG.5-37   Pre-tensioning construction

1—Stretching bed; 2   Beam; 3—Countertop; 4—Prestressing tendon; 5—Grip; 6—Member

bed are <u>bearing all prestressed tensile forces</u>.

The stretching bed <u>consists of</u> countertop, beam and bearing structure. According to the different bearing structure, the stretching bed is divided into a pier pedestal, a trough pedestal, and a pile pedestal.

- *Tensioning machine and grip*:

(1) An anchoring grip for anchoring a prestressing reinforcement to a stretching bed or steel formwork;

(2) The grip for holding the prestressing reinforcement when tensioning.

➢   **Pre-tensioning construction process**

- *Tensioning of prestressed reinforcement*: When multiple sets of tension are carried out, the initial stress of each prestressed reinforcement should be adjusted to make the length and elasticity consistent, so as to ensure the stress of each prestressed tendon after tensioning.

- *Pouring and curing of concrete*:

When determining <u>the mix ratio of prestressed concrete</u>, the shrinkage and creep of the concrete should be minimized to reduce the prestress loss. *When the concrete is poured*, the vibrator must not collide with the prestressing reinforcement. It is also not allowed to collide with the prestressed reinforcement before the concrete reaches the strength.

Concrete can be cured by natural curing or wet and heat curing. When the prestressed concrete members are subjected to wet and heat curing on the stretching bed, a proper maintenance system should be adopted to reduce the prestress loss due to the temperature difference.

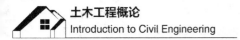

- ***Relaxation of prestressed reinforcement*:**

The prestressing reinforcement can be relaxed only after the concrete strength has reached the value specified by the design. This is because the relaxation prematurely causes a large prestress loss due to the retraction of the prestressing reinforcement. The relaxation of the prestressed reinforcement should be based on the condition and quantity of the reinforcement and the correct method and sequence.

5.5.3 Post-tensioning construction

Construction process

(1) In the fabrication of the members, a hole is reserved in advance at the place where the prestressing reinforcement is placed.

(2) After the concrete reaches the specified strength, the prestressing reinforcement are inserted into the hole and the prestressing reinforcement are clamped by the tensioning device to tension it to the control stress specified by the design.

(3) Anchoring the prestressing reinforcement to the end of the member with the aid of an anchor.

(4) Hole grouting.

➢ **Anchoring and prestressing reinforcement**

Post-tensioned prestressing reinforcement, anchors and tensioning machine are matched, and different types of prestressing reinforcement are used with different anchors.

- ***Type of anchor*:** Screw end anchors, pier anchors, tapered screw anchors, steel cone anchors.

- ***Production of prestressed reinforcement*:** The fabrication of *single thick reinforcement* includes batching, butt welding, cold-drawn and other process. The fabrication of *steel bars* includes straightening, cutting, binding and anchoring.

➢ **Tensioning equipment**

- ***Compose*:** Hydraulic jack, high pressure oil pump and external oil pipe.

Hydraulic jack are driven by high pressure oil pumps to complete the tensioning and anchoring of the prestressing reinforcement.

The high pressure oil pump supplies oil to each cylinder of the hydraulic jack and makes the piston extend or retraction at a certain speed.

➢ **Post-tensioning construction process**

- ***Hole reserved***

Hole reserved is a key process in the post-tensioning construction.

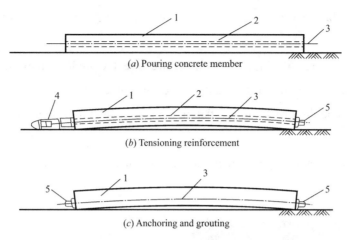

(*a*) Pouring concrete member

(*b*) Tensioning reinforcement

(*c*) Anchoring and grouting

FIG.5-38　Post-tensioning construction

1—Concrete member; 2—Reserved hole; 3—Prestressing tendon; 4—Hoisting jack; 5—Anchor

The hole reserved method includes ***steel tube core-pulling method***, ***rubber tube core-pulling method***, and ***embedded bellows piping method***.

The embedded bellows piping method is only used for curved holes. At the same time as the tunnel is reserved, a grouting hole is reserved at the specified position.

- ***Tensioning of prestressing reinforcement***:

When tensioning the prestressed reinforcement, *the strength of the concrete* should be in accordance with the design regulations. If there are no regulations in the design, it should not be less than 75% of the standard strength of concrete.

After the prestressed reinforcement is tensioned, the hole grouting should be carried out immediately to *prevent corrosion* and increase the *crack resistance* and *durability* of the structure.

- ***Hole grouting***:

Water should be used to flush and moisten the hole before grouting.

During grouting, cement slurry should be injected evenly and slowly without interruption.

Grouting sequence should be first down and then up. Cement slurry should be injected from the lowest point for grouting of curved hole.

5.6　Installation Project

The *structural installation project* is to use the hoisting machinery to assemble the prefabricated components on site according to the drawings to form the main construction process of a complete building.

When formulating a structural installation plan, the following aspects should be solved:

(1) Preparation before the installation of the structure, the production of the components;

(2) Reasonable selection of lifting and transport machinery;

(3) To determine the structural installation method and the installation process of the components;

(4) Determine the hoisting equipment layout and route as well as the site layout of the components.

FIG.5-39 Installation Construction

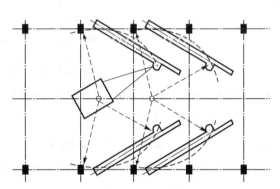

FIG.5-40 Diagram of column hoisting

5.6.1 Hoisting machinery

Hoisting machinery is the key equipment for installation project.

- **Type:** Mast cranes, self-propelled cranes and tower cranes.

The first two categories are mostly used for structural installation of single-storey industrial plants, while the latter are used for structural installation of multi-storey or high-rise buildings.

FIG.5-41 Mast crane

FIG.5-42 Mobile crane

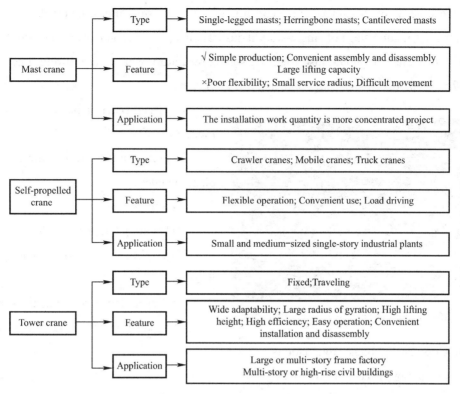

FIG.5-43 Type of hoisting equipment

The choice of crane includes both the choice of *type* and the choice of *model*. *The type of machine* should be selected according to the type of building, the size of the building, and the specific conditions of the installation site. *The mechanical model* should be selected according to the size, weight and position of the mounting member.

5.6.2 Single-story industrial plant installation

The *load-bearing structure* of a single-story industrial plant consists of **foundation**, **column, crane beam, roof truss, skylight truss,** and **roof board.**

Generally, the load-bearing structures of small and medium-sized single-story industrial buildings are mostly assembled reinforced concrete structures. Except for the *foundation pouring* at the construction site, *other components* are mostly made of reinforced concrete **prefabricated** components and transported to the site for lifting. Therefore, structural hoist-

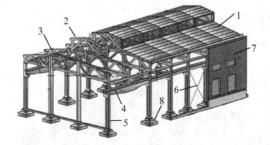

FIG.5-44 Single-story industrial plant

1—Roof board; 2—Skylight truss; 3—Roof truss; 4—Crane beam;
5—Column; 6—Column-bracing; 7—Wall; 8—Foundation

ing has become an important part of the construction of single-story industrial plants.

In the installation stage of single-story industrial plant, in addition to *selecting suitable construction machinery*, it should also focus on some preparatory work before lifting, like *component lifting method, crane driving route* and *layout of members*.

FIG.5-45　Single-story industrial plant installation

➢ **Preparation before hoisting of members**

Preparations for the installation of the plant include:

(1) Site grading;

(2) Building temporary roads;

(3) Laying water and electricity pipelines;

(4) Preparation of the sling and spreader;

(5) Production and layout of members;

(6) Copy the flat line of foundations.

➢ **Hoisting process of members**

The hoisting process of prefabricated components of single-story industrial buildings includes binding, lifting, alignment, temporarily fixing, correction, and final fixing.

● *Hoisting of the column*:

The *binding* of the column should be simple, reliable and easy to install.

The *lifting point* is selected in the lower part of the cow corner, which is higher than the center of gravity of the component and easy to bind.

The column is suspended from the prefabricated position to the cup-like foundation, often using a *rotating method* and a *sliding method*.

FIG.5-46　Fixing of the column

FIG.5-47　Hoisting of the column

The accuracy and verticality of the placement of the column must be strictly corrected.

- ***Hoisting of the beam:***

The crane beam is generally lifted by *two points binding*, aiming at the axis of the top of the corbelr.

The *correction* of the crane beam is mostly carried out after the roof is hoisted.

The crane beam correction includes plane position, verticality and elevation.

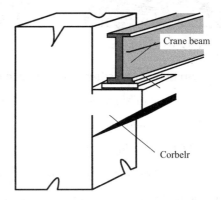

Crane beam

Corbelr

FIG.5-48   Hoisting of beam   **FIG.5-49**   Diagram of crane beam

- ***Hoisting of the roof truss:***

The *binding point* of the roof truss should be selected at the top chord node of the roof truss and symmetrical.

The *angle* between the catenary wire and the horizontal line should not be less than 45° .

The *number and location of lifting points* are related to the form and span of the roof truss.

When the *span of the roof truss* is within 18m, two-point binding is adopted. When the

FIG.5-50   Hoisting of roof truss

span of the roof truss is more than 18m, four-point binding is adopted. The roof truss mainly *corrects* the verticality.

- ***Hoisting of the roof board:***

The roof board is lifted at *four points*.

The *order of hoisting* of the roof board shall be symmetrically installed to the roof edge of the roof truss. After the roof board is in place, it shall be fixed by *welding* at least three fulcrums points. The roof truss should be strictly avoided to bear half load.

CHAPTER 5

FIG.5-51 Hoisting of roof board   **FIG.5-52** Welding of roof board

> **Structural hoisting sequence and crane driving route**

The hoisting of single-story industrial plants can be carried out in two different ways: *partial hoisting* and *comprehensive hoisting*.

● *Partial hoisting*:

The partial hoisting method is carried out in ***two stages***:

In the first stage, the columns in the factory were hoisted at first. The second time, the roof truss was erected and all the crane beams, connecting beams and column-bracing were hoisted.

In the second stage, the roof system is hoisted, that is, the roof truss, the skylight truss, the roof board and the roof bracing are hoisted for the third time of the crane set off.

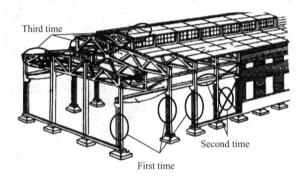

FIG.5-53 Partial hoisting

Feature: The *spreader* does not need to be changed frequently. The *operation procedures* are basically the same, and the *crane* can be fully utilized and the installation speed is fast. *However*, this installation method cannot provide a working surface early for the following work, and the crane travels long.

● *Comprehensive hoisting*:

The *comprehensive hoisting* method refers to the crane in a trip to install all the members. At first install 4 to 6 columns, and immediately correct and final fix columns; then install the crane beam, connecting beam, roof truss, roof board and other members. After the lifting of one section is completed, the crane moves to the next section for further installation.

Feature: This type of installation method is characterized by *fewer parking points* and *short driving routes*, which can create a working surface for subsequent work such as equipment installation. *However*, the crane cannot achieve maximum efficiency, the installation speed is slow, and the correction and fixing time of the members are urgent.

5.6.3 Multi-story assembled frame structure installation

The multi-story assembled frame structure has *small plane size* and *high height*, and the *types* and *numbers* of members *are large*, the *joints are complicated*, and the *technical requirements are high*. Therefore, when considering the installation plan, the problems of **selection and arrangement of the hoisting machinery**, **installation sequence** and **installation method** should be addressed.

> ➤ **Site preparation before installation**

The preparation work before the installation of the members mainly ***includes*** the copy the flat line, the inspection of the members and the snap the line, the discharge of the members in place and the preparation of the foundation. The *preparation of members* is mainly transportation, stacking, inspection, and snapping the line.

The inspection and preparation of the members is to check the model, size and appearance quality of the members, and the members are discharged in place. At the same time, the order of hoisting should be considered to facilitate the locating of member.

The arrangement of the members shall be based on the site conditions, using the components parallel to the crane track, perpendicular to the track or oblique to the track.

> ➤ **Hoisting of members**

The prefabricated frame structure hoisting is mainly the installation of the prefabricated columns, beams, slabs and stairs and their node processing.

- ***Hoisting sequence of members:*** *partial hoisting* and *comprehensive hoisting*.
- ***Hoisting process of members:***

When the *column* is hoisted, it is mainly for the protection of the joint extension steel bar, so that the welding of the steel bar after the lifting is aligned. When the column is hoisted in place, the axis should be aligned and the column vertical.

Installation of *beam and slab* must be carried out after the concrete at the lower end of the column reaches the required strength. The slab is generally hoisted after the beam is installed and corrected and fixed.

5.7 Decoration Projects

The *purpose* of the decoration project is to improve the insulation, sound insulation and moisture resistance, improve the living conditions and increase the aesthetics of the building. The construction *features* of the decoration project are large engineering volume, long construction period, large amount of labor and high cost.

Decoration works is the last construction process of the project, generally includes *plastering engineering*, *facing engineering*, *painting engineering* and *ceiling engineering*, *doors* and *windows* and *curtain wall engineering*.

FIG.5-54 Installation of multi-story frame structure

FIG.5-55 Decoration Engineering

5.7.1 Plastering engineering

Plastering engineering are divided into general plastering and decorative plastering.

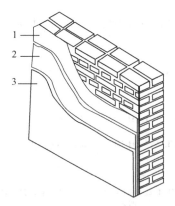

FIG.5-56 Plaster composition
1—Bottom layer; 2—Middle layer; 3—Top laye

FIG.5-57 General plastering construction

The plaster layer is generally composed of *a bottom layer*, *a middle layer*, and *a top layer*. The role of the <u>bottom</u> mortar is to enhance the combination with the base structure, the <u>middle</u> mortar plays a leveling role, and the <u>surface</u> mortar ash mainly plays a decorative and smoothing role. *The type of mortar* used in general plastering should be determined according to the type of plaster base and the location and environment of the plaster layer.

➤ **General plastering construction**

- *Type*: Ordinary plastering, intermediate plastering and advanced plastering.
- *Construction sequence*: First outdoors and then indoors;

  First above and below;

  First ground and then the top of the wall.

- *Preparation*:

The base layer is treated to ensure a firm bond between the plaster layer and the base layer.

Check the flatness and verticality of the wall to determine the total thickness of the plaster layer.

- *Construction process*: The bottom layer should be firmly bonded. The middle layer should be flattened and compacted to achieve compactness, smoothness and roughness. The surface layer is compacted, flattened and calendered according to different conditions.

➤ **Decorative plastering construction**

- *Definition*: Decorative plastering *refers to* the surface layer of the wall surface with different color effects by using the material characteristics and processing.
- *Plastering requirements*: *The bottom layer* of the decorative plaster is the same as the general plastering requirement, except that *the surface layer* has different forms depending on the material and the construction method.
- *Construction process*: In the common decorative plastering layer, the water brush stone and the sputum stone are the same except for the surface treatment methods, while the spraying, rolling, brushing and smearing methods are different except for the mortar. The bottom layer and middle layer practices are similar.

5.7.2 Facing engineering

- *Types*: Natural stone facing, decorative brick facing, artificial stone facing, metal facing and wooden facing.
- *Requirements*: The base layer should have sufficient stability and rigidity, and the surface should be smooth, clean, rough and suitable humidity to ensure the quality of the decorative surface.

FIG.5-58　Natural stone facing

FIG.5-59　Artificial stone facing

● *Feature*:

The natural stone facing is installed on the inner and outer walls and cylinders of the building, which can effectively improve the decorative effect and artistic quality of the building and its space environment.

The artificial stone facing is light weight and high strength, rich in color and safe to use.

The decorative bricks facing are divided into exterior and interior facing tiles. The inner wall is generally made of ceramic tile, and the exterior wall facing tile has ceramic mosaic tile and glass mosaic.

The metal facing feels good, simple and straight. The most common is the metal exterior wallboard, which has the advantages of elegant solemnity, firmness, light weight, durability and easy disassembly.

FIG.5-60　Decorative brick facing

FIG.5-61　Metal decorative surface

5.7.3　Ceiling engineering

The ceiling is the upper interface in the hexahedron of the interior space and is the most varied part.

- ***Types***: Direct ceiling, suspended ceiling.

Direct canopy refers to directly plastering, spraying or pasting other decorative materials on the bottom of the floor. It is suitable for interior decoration projects with low requirements.

Suspended ceiling refers to the keel and the cover decorative layer is suspended under the floor by using the boom. In the interlayer space of the floor and the cover layer, air conditioning, fire fighting, lighting and other pipelines can be arranged, and thermal insulation materials can also be placed.

FIG.5-62   Direct ceiling   **FIG.5-63**   Suspended ceiling

- ***Keel construction process***: The construction process of light steel keel is snap the line, fixed boom, installation main keel, leveling main keel, fixed secondary keel, installing cross bracing keel, and covering panel.

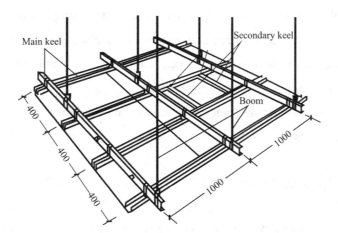

FIG.5-64   Diagram of keel structure

5.7.4   Curtain wall engineering

- ***Types***: metal-framed curtain wall; frameless curtain wall and point glass curtain wall.

Metal-frame type curtain wall is divided into bright frame glass curtain wall, semi-hidden frame glass curtain wall and hidden frame glass curtain wall.

Frameless curtain wall is divided into glass frame and no glass frame curtain wall.

Point glass curtain wall is divided into four point type (x-shaped h-shaped), three-point type (y-shaped), two-point type (v-shaped and 1-shaped and single-point type, etc.).

FIG.5-65 Metal-frame curtain wall   **FIG.5-66** Point glass curtain wall

- *Requirements*:

The structural part installed on the glass curtain wall shall comply with the provisions and design requirements of the relevant structural construction and acceptance specifications.

In order to make the construction of the glass curtain wall safe and smooth, a separate construction organization design plan should be prepared.

- *Construction process*: snapping the line, installing connecting iron parts, installing columns and beams.

5.8 Steel Structure Construction

In a broad sense, *steel structure engineering refers to* a structure in which steel is used as a base material and mechanically assembled. Due to its high strength, light weight, short construction period and high precision, steel structures are widely used in civil engineering such as buildings and bridges.

5.8.1 Steel processing technology

➢ **Lofting, marking and baiting**

Lofting is the process of drawing the real shape of the product on the lofting table or plate according to the shape and size of the product construction detail according to the ratio of 1 ∶ 1, and obtaining the actual length and making the sample.

Marking is based on the sample of steel on the steel components and marked with a variety of processing marks, for the steel cutting preparation.

The purpose of the *baiting* is to separate the shape of lofting and marking parts from the raw material.

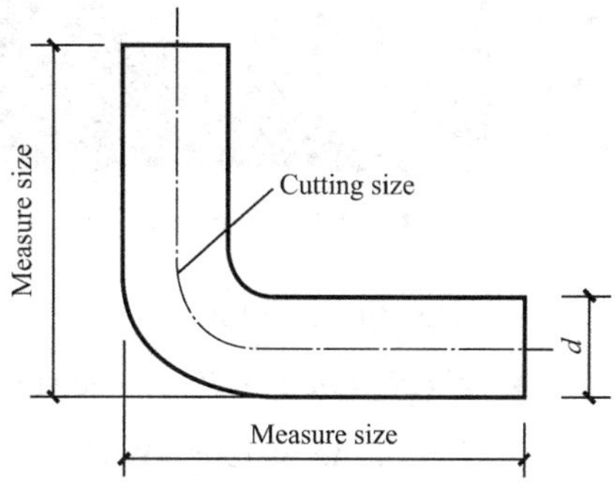

FIG.5-67   Lofting

➢ **Component processing**

✓ Correction

Before the steel is used, the steel material will be deformed due to *remaining stress* inside the material and improper storage, transportation and lifting. In order to ensure the quality of the steel structure and the installation, *it is necessary to correct the components that do not meet the technical standards*.

- *Definition*: The correction of the steel structure is the process that meets the requirements of the steel and meets the technical standards by external force or heating.
- *Types*: straightening; flattening and shape righting.
- *Methods*: flame correction; mechine correction and manual correction.

(1) Flame correction

The flame correction of steel is accomplished by using flame to locally heat the steel, so that the heated metal will produce compression plastic deformation due to the expansion obstruction, and the longer metal fiber will shorten after cooling.

(2) Mechanical correction

The essence of mechanical correction is to make the bent steel produce excessive plastic deformation under the action of external force to achieve the purpose of straightness. The advantages are high force, low labor intensity and high efficiency.

(3) Manual correction

The manual correction of steel is carried out by hammering, and the operation is simple and flexible. Manual correction is used to correct small-sized steels due to small correction force, high labor intensity, and low efficiency.

FIG.5-68 Flame correction   **FIG.5-69** Mechanical correction

✓   Bending and forming

● ***Steel plate hending***: The steel plate bending is formed by continuous three-point bending of the material by a rotating roller shaft. Steel plate bending process includes three processes: prebending, neutralizing and bending.

● ***Profile steel***: Bending of shape steel, bending of steel tube.

When the **shape steel** is bent, the gravity center line of the section and the action line of the force are not in the same plane, so the section will deform. The degree of deformation depends on the stress, which in turn depends on the bending radius. *The bending radius is smaller, the deformation degree is greater.* In order to control stress and deformation, *the minimum bending radius* should be controlled.

FIG.5-70 Steel plate bending   **FIG.5-71** Bending of shape steel

When the **steel tube** is bent under the action of external forces, its section will be deformed, and *the outer tube wall will be thinner, while the inner tube wall will be thicker*. During the bending process, the deformation of the steel tube can be reduced by adding filler (sand or spring) to the steel tube.

● ***Edge preparation***: In the manufacture of steel structures, the internal structure of the sheared or gas-cut steel plate is hardened and metamorphosed. In order to ensure the quality of heavy components such as bridges or heavy crane beams, the edges need to be machined.

✓ *Other processes*: folding, mold pressing, hole making.

FIG.5-72 Folding   **FIG.5-73** Hole making

In the process of steel structure manufacturing, the operation process of bending the member's edge into a dip Angle or a certain shape is called *folding*. *Mould pressing* is a process of forming steel with mould on pressure equipment. *Hole making* is the process of machining holes in steel.

5.8.2 Assembly and connection of steel structures

➤ **Factory assembly**

Due to the restrictions of transportation, lifting and other conditions, sometimes the members are divided into several sections. In order to ensure the smooth installation, the *pre-assembly* should be carried out before leaving the factory according to the design requirements of the members.

In the pre-assembly, in addition to *checking the size* of each part of the bolted joint plate, *the pass rate* of the hole of plate should also be checked. After the pre-assembly inspection is passed, *the upper and lower positioning center lines*, *the elevation reference line*, and *the intersection center point* should be clearly and accurately marked.

➤ **Welding construction method**

● *Definition*: Welding is a process in which the parts to be connected are heated to a molten state and then connected.

● *Advantages*: No drilling is required, the structure is simple, the processing is easy, and the cross section of the member is not weakened.

- *General requirements for welding*:

(1) The selection of the welding method shall take into account the material and thickness of the welded component, the form of the joint and the welding equipment;

(2) Efficiency and economy of welding;

(3) Stability of welding quality.

- *Welding methods and characteristics*: Manual welding, semi-automatic welding and automatic welding.

FIG.5-74   Manual welding

FIG.5-75   Automatic welding

Automatic welding has high production efficiency and good welding quality, and is suitable for long butt welding or fillet welding. *Semi-automatic welding* is more flexible than automatic welding and is suitable for short length and curved welds.

- *Welded joint form*: Butt joints, fillet joints, T-joints and lap joints.

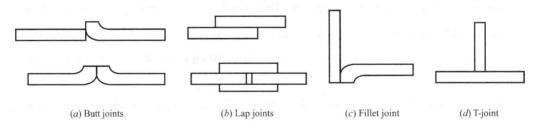

(*a*) Butt joints   (*b*) Lap joints   (*c*) Fillet joint   (*d*) T-joint

FIG.5-76   Welded joint form

- *Weld form*:

(1) According to the spatial position of welding, the weld form can be divided into three types: horizontal weld, vertical weld and overhead weld.

(2) According to the combination form, the weld can be divided into three types: butt weld, fillet weld and plug weld.

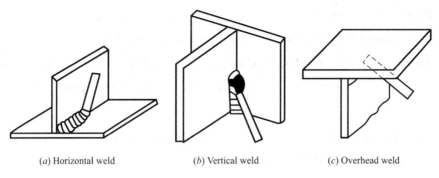

(a) Horizontal weld   (b) Vertical weld   (c) Overhead weld

FIG.5-77   Weld form

- *Selection of welding process parameters*: core diameter, welding current, arc voltage, welding layer.

➢ **Bolted construction**

- *Ordinary bolts*: Ordinary bolts are one of the commonly used fasteners for steel structures. They are used to connect and fix components between steel structures, or to fix steel structures to the foundation to make them a whole.

Common bolts commonly used are *hexagon bolts*, *stud bolts* and *anchor bolts*.

- *High-strength bolts*: High-strength bolts are special bolts made of high-quality carbon steel or low-alloy steel, which have high strength.

FIG.5-78   Bolted connection   **FIG.5-79**   Bolted connection of column base

FIG.5-80   Hexagon bolt   **FIG.5-81**   Stud bolt   **FIG.5-82**   Anchor bolt

✓ **Types**: High-strength bolts are divided into three types according to the connection form: *tension connection, friction connection* and *pressure connection*.

✓ **Advantages**: The installation is simple and rapid, the assembly and disassembly is convenient, the bearing capacity is high, the force performance is good, and the safety is reliable.

✓ **Application**: Important structures such as large-span structures, industrial plants, bridge structures, and high-rise steel frame structures.

5.9 Development Prospect of Construction Technology

➢ **Development direction of *construction management informatization***

With the development of information technology, it has become an inevitable development trend to realize the construction information management of construction enterprises. Applying ***Internet technology*** to construction <u>remote monitoring</u>, <u>strengthening supervision</u> and <u>control of construction process</u> is a new direction of construction management development.

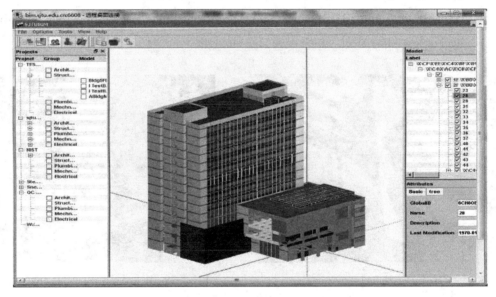

FIG.5-83 Building Information Model

In the aspect of ***construction process control***, computer aided function plays an important role in <u>optimizing construction plan</u>, such as ***CAD drawing scheme design of formwork and scaffold, automatic mixing control of concrete, measurement of mass concrete temperature, a large number of data collection and collation***. Computer technology will also give full play to its advantages in more detail in construction management.

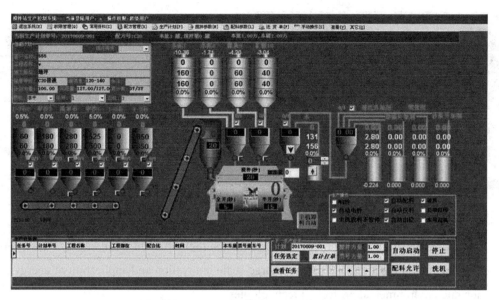

FIG.5-84   Automatic mixing control of concrete

> **Development direction of *construction technology innovation***

As the scale of large-scale buildings continues to expand, the overall structure will be more complicated. Therefore, ***informatization***, ***greening*** and ***automation*** for large-scale construction will inevitably become the innovative direction of construction technology.

From the current situation of China's construction industry and the application of construction technology, the future of China's construction should be the ***large span steel structure*** construction, ***super high-rise building*** construction, ***deep foundation*** construction, ***building renovation***, ***bridge*** construction, ***information*** construction, ***environmental protection*** construction and other aspects of construction technology as the focus of research and innovation.

It is an inevitable trend to improve the modernization degree of construction technology by ***using mechanical automation construction technology*** to replace part of manual construction technology, ***using refined construction technology*** instead of extensive construction technology, and ***using green construction technology*** to replace high energy consumption and heavy pollution construction technology.

> **Development direction of *energy conservation and environmental protection* of buildings**

The construction industry has always been regarded as a traditional industry with <u>high energy consumption and high pollution</u>. In the context of building an ecological country, the construction industry also needs to achieve a shift to ***energy conservation and environmental protection***, while paying attention to ***economic benefits*** while paying attention to ***social benefits***.

FIG.5-85   Environmental protection construction

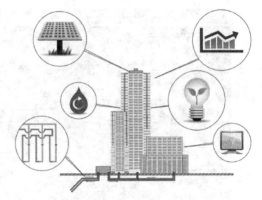

FIG.5-86   Green building

Therefore, construction technology must attract sufficient attention in energy conservation and environmental protection, and promote large-scale innovation and application of energy conservation and environmental protection technologies.

> ➤ *New technologies in other industries* **will be more widely applied to construction**

For example, the first ***dynamic spray technology*** for agriculture, the ***thermochromic paint*** for the transportation industry, the ***smart windows*** for energy-saving color light systems, etc., have been widely used in the construction of buildings, and there will be more non-construction industries technological innovation is applied to construction in the future.

Although China has made certain achievements in building construction technology, it is far from enough for the rapid development of urban modernization. To this end, we must continue to carry out technological innovations on the basis of maintaining the current state of construction technology, open up new areas of technology, and self-improve by drawing on advanced foreign technology to adapt to the needs of the development of construction projects.

Questions

- Briefly describe the characteristics of various construction machinery in earthworks.
- Draw a schematic diagram of the sinking foundation and briefly describe the role of each part.
- Describe the construction process of masonry construction.
- What are the forms of the template? What are the requirements for the template to be removed and installed?
- Describe the construction process of the pre-tensioning method and the post-tensioning method.

- What are the methods for hoisting a single-storey industrial plant? Briefly describe the specific steps of each method.
- What are the connection methods of steel structures?
- What direction do you think future construction technology will develop?

Reference List

[1]   Rao Bo. Building construction technology [M]. Shanghai: Tongji University Press, 1997.

[2]   Zhu Yan, Liang Shaozhou, Zhang Yurong . Building Construction: Building Construction Technology[M].Beijing: Tsinghua University Press, 1994.

[3]   Ying Huiqing. Civil Engineering Construction[M]. Shanghai: Tongji University Press, 2001.

[4]   Yu Guofeng. Civil Engineering Construction Technology [M]. Shanghai: Tongji University Press, 2007.

[5]   Ren Jiliang, Zhang Fucheng, Tian Lin. Building Construction Technology [M]. Beijing: Tsinghua University Press, 2002.

CHAPTER 6
CONSTRUCTION MANAGEMENT

6.1 Construction Procedures and Construction Codes

6.1.1 Procedures of construction

The construction procedures, summary of essential construct works, is the order that each work must be followed and the centralized reflections of the fixed objective regular in the procedures of construction. The construction procedure of engineering projects can generally be divided into five stages: planing and decision stage, design stage, construct prepare stage, construction and pre-use preparation and completion acceptance. As shown in FIG.6-1.

- *Planning and decision stage*

Planning and decision stage major includes recommendation of project and feasibility study.

(1) Recommendation of project

According to the intention of the main investment, consultation investigate and analysis the investment opportunities, then write proposal documents and hand in it to relevant departments that set by the nation.

(2) Feasibility study

A verify work to the technical and economic rationality of a proposed project. If the feasibility study don't be approved, the next step can not be carried out.

- *Design stage*

(1) General project

Preliminary design and construction drawing design.

(2) Special project

Preliminary design, technical design, construction drawing design.

- *Construction prepare stage*

Land requisition, remove, flatting site, provide water, electricity and road, material

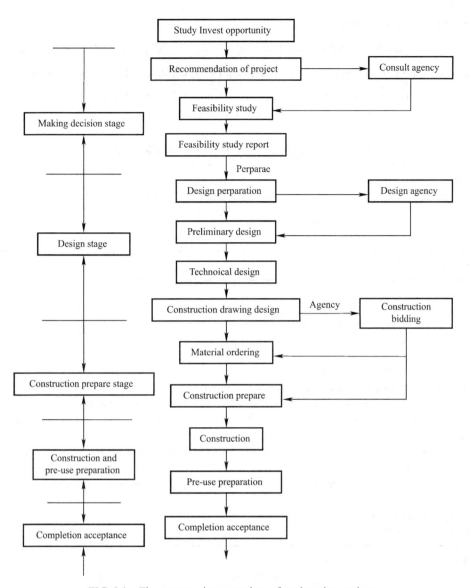

FIG.6-1   The construction procedure of engineering projects

ordering, organizing construct bidding, selecting construct units, application of start-up reports, etc.

- ***Construction and pre-use preparation***

(1) Constructor

Finish the construction main body and installation according to the design document.

(2) Owner

With the assistance of the construction unit to do the series of preparation job to project use,
such as personnel training, organizational preparation, technical preparation and material preparation, etc.

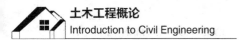

• *Completion acceptance*

Application for acceptance needs to arrange technical document, draw up project completion drawings, prepare project final accounts, etc. The acceptance of the project means delivered to use, at the same time stipulating the real-time warranty.Engineering quality acceptance report as shown in Table 6-1.

Table 6-1   Engineering Quality Acceptance Report

Unit project's name			
Construction area		Structure type and layers	
Construction unit's name			
Construction unit's address			
Construction unit's P.C		TEL	

Quality acceptance advice:(The following content should be included)

Completion requirements:

Fill requirements:

1. Implementation of quality responsibility of construction units.

2. Whether the project has been completed as contents of the contract.

3. Performance of construction units conduct mandatory standards and principles during construction.

4. During the construction process, whether the quality problem of the request of the supervision has been corrected and approved by the supervision unit.

5. After the completion of the project, the enterprise self-check whether the project meet the completion standards, structural safety and functional requirements.

6. Engineering quality assurance information (including functional test reports) are basically complete and have been bound as required.

7. Building settlement observation results and tilt rate.

8. Other information need to be explained.

Project Manager: Date Month Year	
Head of enterprise (chief of Quality): Date Month Year	Seal of Construction Unit
Head of enterprise technology (chief engineer): Date Month Year	
Chief executive Officer : Date Month Year	

(1) Large or medium-sized projects:

Initial check and completion acceptance.

(2) Simple or small-sized projects:

Only includes completion acceptance.

6.1.2 Construction laws & codes

- *Definition*

The general term of laws and regulations aimed at adjusting the various social relations among the organs, enterprises, institutions, social organizations with citizens in construction activities. It formulated by a national legislature or a approved administrative.

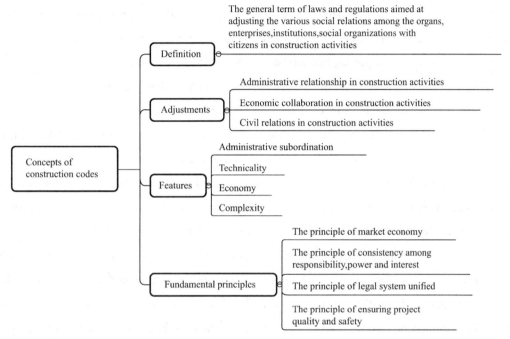

FIG.6-2 Construction Laws & Codes

- *Adjustment*

All kinds of construction relations forming and embodying the rights and obligations of both parties in the course of construction activities.It includes three parts: Administrative relationship in construction activities, Economic collaboration in construction activities, Civil relations in construction activities.

(1) Administrative relationship in construction activities: The corresponding management relationship among the nation and its build administrative departments, the construction units, design units, and related units (such as intermediary service agencies). It includes two related aspects; One is planning, guidance, coordination and services; The other is inspection, supervision, control and mediation.

(2) Economic collaboration in construction activities: A horizontal cooperative relationship

of equality, voluntary, mutual benefit and assistance. such as the survey, design and construction relationship between the construction unit and the survey design unit, etc.

(3) Civil relations in construction activities: The relationship of civil rights and obligations between states, legal persons of enterprise and citizens arising out of construction activities.

- *Feature*

In addition to the basic features of general laws, construction codes also have the features of administrative subordination, technicality, economy and complexity.

(1) Administrative subordination: It is the main feature of the construction codes, which determines that the construction codes must adopt the adjustment mode that directly reflects the activities of administrative power, and adjust the legal relationship of the construction activities by the way of administrative instructions. Adjustment mode includes authorization, command, prohibition, permission, exemption, confirmation, plan and revocation.

(2) Technicality: A large number of construction codes take the form of technical specifications, such as various design standard, construction standard, acceptance standard, etc.

(3) Economy: The Construction activities are closely linked to production, distribution, exchange and consumption, and directly creating wealth for the society and increasing the accumulation of the country. Such as real estate development, residential commercialization, etc.

(4) Complexity: The construction legal norms take the relationship with the rights and obligations as the object of adjustment, including the construction administrative legal relationship between unequal subjects, the construction civil legal relationship between equal subjects, and the construction technology legal relationship between the two. The construction laws and regulations take the actors engaged in construction activities as the objects of adjustment, including the state, citizens, legal persons and social and economic organizations.

- *Fundamental principles*

The principles of construction codes include the principle of market economy, the principle of consistency among responsibility, power and interest, the principle of legal system unified and the principle of ensuring project quality and safety.

6.1.3　System of construction codes

The basic framework of the construction codes system consists of vertical structure and horizontal structure.

- *Vertical structure of the Construction codes system*

According to the provisions on legislative permissions *Legislative Law of the People's Re-*

public of China (《中华人民共和国立法法》) and the requirements of *Plan of Construction codes system* (《建设法规体系规划方案》), China's construction law system is defined as the longitudinal form of trapezoidal structure. As shown in **FIG.6-3**.

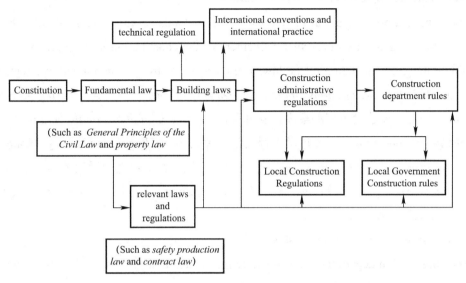

FIG.6-3 Vertical structure of the Construction codes system

(1) Constitution

The Constitution is the fundamental law of the state and has the highest legal status and effect, and any other laws must conform to and not conflict with it.It is the legislative basis for the engineering construction. At the same time, it clearly stipulates the principles of the national fundamental construction, and directly regulates and adjusts the related activities of the project.

(2) Fundamental law

The basic law is formulated and amended by the National People's Congress, stipulating and regulating a certain aspect of national and social life relations with basic and comprehensive laws, such as *General Principles of the Civil Law of the People's Republic of China* (《中华人民共和国民法通则》), *Tendering Law of the People's Republic of China* (《中华人民共和国招标投标法》), *Contract Law of the People's Republic of China* (《中华人民共和国合同法》), etc..The legal effect of the Basic Law is lower than Constitution, but higher than other laws.

(3) Building laws

The Building laws is enacted and promulgate by the National People's Congress or the Standing Committee of the National People's Congress, is the core and foundation of the system of construction laws, belongs to the construction administrative department's business scope, such as *Building Law of the People's Republic of China* (《中华人民共和国建筑法》), *Urban and Rural Planning Law of the People's Republic of China* (《中华人民共和国城乡规划法》) and *Urban Real Estate Administration Law of the People's Republic of China* (《中华人民共和国城市房地产管理法》),*etc.*

(4) Construction administrative regulations

The construction administrative regulations are formulated and promulgated by the State Council, which is generally the further refinement of the construction legal provisions in order to facilitate the implementation of the law. The current construction administrative regulations mainly mainly include *Regulations on the Management of Construction Engineering Reconnaissance and Design* (《建设工程勘察设计管理条例》), *The regulations on the quality Management of Construction projects* (《建设工程质量管理条例》), *Regulations on the Management of Safety in production of Construction projects* (《建设工程安全生产管理条例》), *and Regulations on the Management of Urban Real Estate Development and Management* (《城市房地产开发经营管理条例》), and so on.

(5) Construction department rules

The construction department rules shall be formulated and promulgated by the Ministry of Housing and Urban and Rural Development or other departments under the State Council, such as *Measures for Construction and Bidding for Construction of Construction Projects* (《工程建设项目施工招标投标办法》), promulgated by the seven ministries, and *Measures for Quality Management of Projects Reconnaissance* (《建设工程勘查质量管理办法》) issued by the Ministry of Housing and Urban and Rural Development, *Regulations on Qualification Management of Construction Enterprises* (《建筑业企业资质管理规定》), and so on.

(6) Local construction regulations

The people's congresses and their standing committees of provinces, autonomous regions and municipalities directly under the Central Government may formulate local construction regulations applicable only to their respective administrative regions on the premise that they do not conflict with different constitutions, laws and administrative regulations in accordance with the specific conditions and actual needs of their respective administrative regions, such as *Regulations on Industrial Injury Insurance in Shaanxi Province* (《陕西省工伤保险条例》).

(7) Local government construction rules

The people's governments of provinces, autonomous regions, municipalities directly under the Central Government and autonomous prefectures with districts may formulate regulations in accordance with laws, administrative regulations and local laws of provinces, autonomous regions and municipalities directly under the Central Government to make Local Government Construction rules, such as *Fire Regulations on the Supervision and Administration of Construc-*

tion Project of Shaanxi Province (《陕西省建设工程消防监督管理规定》), *etc.*

(8) Legal interpretation

when the construction law needs to further clarify the specific meaning or new circumstances arise after the enactment of the law, the Standing Committee of the National People's Congress may make a legal interpretation; The supreme people's court and the supreme people's procuratorate may interpret the specific application of the law in the judge and check work.No judicial organ other than the Supreme People's Court or the Supreme People's Procuratorate may interpret the specific application of the law.

(9) Technical regulation

Technical documents such as technical regulations, standards and quota methods that are formulated or approved by the state and are effective throughout the country.They are the basis for construction engineering technicians to engage in economic operations and building management monitoring, such as budget quota, design code, construction specification, acceptance standard, etc.

(10) International conventions and international practice

International conventions refer to bilateral and multilateral conventions, agreements and other documents with the nature of conventions concluded, participated in, signed, joined and recognized by China and foreign countries. The names of international conventions, in addition to conventions, include treaties, agreements, protocols, charters, covenants, exchange of letters and joint declarations.

International practice refers to the rules of international law confirmed and embodied by various international adjudicators and some unwritten habits in international exchanges.

- *Horizontal structure of the construction codes system*

According to *Plan of Construction codes system* (《建设法律体系规划方案》) formulated by the construction administration department under the state council, eight laws have been enacted and three have been implemented. They are *Urban and Rural Planning Act* (《城乡规划法》), *Urban Real Estate Management Act* (《城市房地产管理法》) and *Building Act* (《建筑法》). As shown in FIG.6-4.

(1) Urban and rural planning act

The law of urban and rural planning is a general term for the legal norms to regulating the social relations that occur during the construction of various projects in the planning area. The purpose of the legislation is to determine the size and development direction of the city, to achieve the economic and social development goals of the city, to formulate the city planning rationally and to speed up the construction.The current *Urban and Rural Planning Act* (《城乡规划法》) came into force on 1 January, 2008 .

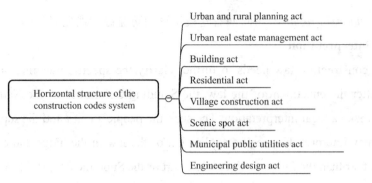

FIG.6-4 Horizontal structure of the Construction codes system

(2) Urban real estate management act

Urban real estate management law is a general term for the legal norms to adjust urban real estate industry and various real estate business activities and their social relations. Its legislative purpose is to protect the legitimate rights and interests of urban real estate operators, real estate owners, users, to promote the development of real estate industry and to meet the needs of socialist modernization and people's lives. *Urban Real Estate Management Act* (《城市房地产管理法》) was promulgated on July 5, 1994 and came into force on January 1, 1995. It was revised twice on August 30, 2007 and August 27, 2009.

(3) Building act

Building Act is a general term for the legal norms regulating construction activities and their social relations.Its legislative purpose is to strengthen the management of construction activities, maintain the order of the construction market, protect the legitimate rights and interests of the parties involved in construction activities, regulate the qualification of the main body of the construction industry, standardize the construction behavior, and ensure the construction quality. *Building Act* (《建筑法》) was promulgated on November 1, 1997 and came into force on March 1, 1998. It was revised on April 22, 2011.

(4) Residential act

Housing Act is a general term of the legal norms regulating the ownership of urban housing, the financing of housing construction, housing management and maintenance and their social relations.Its legislative purpose is to protect citizens' right to enjoy housing, to ensure the legitimate rights and interests of residential owners, to promote the development of housing construction, and to constantly improve citizens' housing conditions and living standards.

(5) Engineering design act

Engineering design law is a general term of adjusting the qualification management, quality management and technical management of engineering design, as well as the whole process of formulating design documents and its social relations.Its legislative purpose is to strengthen the management and improve the level of engineering design.

(6) Municipal public utilities act

Municipal Public Utilities Act is a general term of the legal norms regulating the construction, management activities and social relations of municipal facilities and public utilities, city appearance, environment and sanitation, and landscaping.Its legislative purpose is to strengthen the unified management of municipal public utilities, to ensure the smooth progress of the construction and management of urban public utilities, and to give full play to the multifunctional role of cities.

(7) Scenic spot act

Scenic spot Act is a general term for the legal norms regulating the social relations arising from the activities of protecting, utilizing, developing and managing scenic resources.Its legislative purpose is to strengthen the management, protection, utilization and rational development of scenic spot resources.

(8) Village construction act

Village construction Act is a general term of the legal norms regulating the planning, comprehensive development, design, construction, public infrastructure, housing and environmental management of villages and their social relations.Its legislative purpose is to strengthen the construction management and continuously improve the environment, of villages and towns, promote the coordinated development of urban and rural economy and society, and promote the construction of new socialist villages and towns.

6.1.4 Overview of foreign construction codes

- *American system of construction codes*

In the construction field of the United States, only one *Uniform Building Code*, (UBC,《美国统一建筑条例》) regulates the construction, alteration, relocation, demolition, maintenance, protective use, building administration and permit system of buildings.

Building codes in the United States are autonomous projects, not under the jurisdiction of the central government. They are formulated by the people and become mandatory through the legislative process of the state governments, such as hearing, voting or parliamentary approval. The current U.S. Construction codes is being pushed by three groups:

(1) ICBO (International Conference of Building Officials), adopted by the western states.

(2) BOCA (Building Official and Code Administrators International), adopted by the North-

West and Eastern States.

(3) SBCCI (Southern Building Code Congress International), adopted by the southern states.

At present, ICBO, BOCA and SBCCI are being integrated into ICC (International Code Council Inc).

American construction law mainly includes civil and commercial law, economic law, administrative law and other. The details are as shown in FIG.6-5.

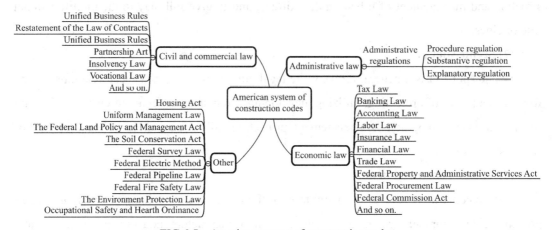

FIG.6-5　American system of construction codes

- ***Eu system of Construction codes***

The laws and regulations of the EU construction market include technical regulations and technical standards.

(1) Technical regulations

Definition: Mandatory requirements for product characteristics or corresponding processing and production methods and applicable administrative regulations.

Feature: Mandatory.

Types: Treaty, directives, and decision.

(2) Technical standards

Definition: The approaches and methods adopted to meet the basic requirements stipulated in the regulations, including technical requirements, testing and other specific measures, shall be formulated into technical standards.

Features: Recommended, Voluntary.

- ***Japan system of Construction codes***

The Japanese construction industry has three basic laws.

(1) Building Standard Act (《建筑基准法》): The minimum standards for the use of land,

structures, equipment and use of the buildings are specified in detail, with an emphasis on the technical aspects.

(2) Japanese Construction Law (《日本建筑业法》): Mainly for the permission of the construction industry, engineering contracting, handling dispute processing has made the detailed regulations, is about the whole construction industry in the stipulations on the management of all kinds of engineering, including the scope of the construction industry more widely than building, not only including civil engineering, but also including civil engineering related industries, such as stone material engineering, pipeline engineering, glass engineering, paint, etc.

(3) Architects Law (《建筑师法》): The basic laws governing the qualifications of architects.

Other major construction-related laws are as shown in the FIG.6-6.

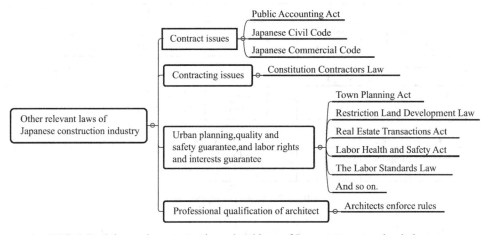

FIG.6-6 Other major construction-related laws of Japanese construction industry

6.2 Project Management

The project management is to take the project as the management object, under the established constraint condition, in order to realize the project goal optimally, according to the project internal law, carries on the effective plan, the organization, the command, the control and the coordination system management activity to the project life cycle whole process.

6.2.1 Composition and classification of projects

● *Definition*:

Under certain constraints, an organized one-time project construction work or task with a complete organization and specific clear goals.

- ***Features***:

Disposable, targeted, restrictive, cyclical life, composed of activities, large investment, long construction cycle, many uncertain factors, large risks, and many participants.

- ***Composition***:

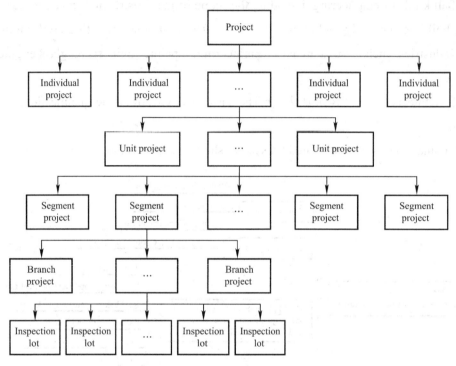

FIG.6-7 Project composition

(1) Individual project: A group of engineering projects with independent design documents that, upon completion, can give full play to production capacity or benefits independently.

(2) Unit project: The project which has independent construction drawings and can organize construction independently, but cannot independently exert production capacity or benefit after completion.

(3) Segment project: A number of divisions that make up a unit project are called segment project.

(4) Branch project: A number of divisions that make up a segment project are called branch project.

(5) Inspection lot: An inspection body consisting of a certain number of samples that are summed up for common inspection under the same production conditions or in a prescribed manner.

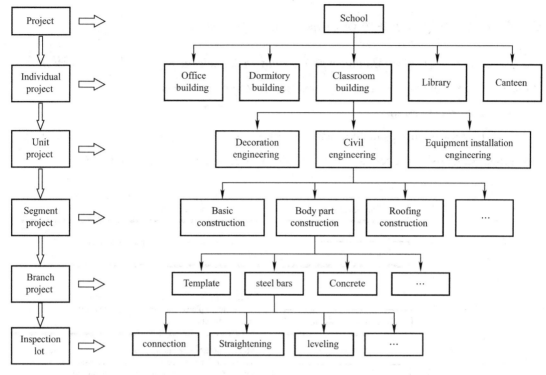

FIG.6-8   Project composition of a school as an example

- *Classification of projects*

Different classification of engineering projects from different perspectives, as shown in Table 6-2.

Table 6-2   Classification of projects

classification standard	content
Feature	New projects, expansion projects, reconstruction projects, recovery projects, relocation projects
Function	Residential buildings, public buildings, industrial buildings, infrastructure
Tasks undertaken by the participants	Engineering projects (including use to scrap), engineering contracting projects, engineering survey and design projects, engineering supervision projects
Project benefit and market demand	Competitive projects, fundamental projects, public welfare projects
Industry composition and application	Productive engineering projects (Details: Agriculture, forestry, animal husbandry, fishery, water conservancy and its services; Industry; Geological survey and exploration; Construction industry; Transportation, posts and telecommunications; Commerce) Non-productive engineering projects (Details: Real estate management, public utilities, residential services and advisory services; Public health, sports and social welfare; Education, culture, arts, radio and television; Scientific research and integrated technical services; Finance; Insurance; State organs, government organs and public organizations, etc.)
Major	Construction project, civil engineering project, line and pipeline installation project, decoration project

6.2.2 Knowledge system of project management

According to 《*Project Management the Body of Knowledge*》(《项目管理知识体系》), Project management has five process groups and ten knowledge areas.

- ● *Five process groups*

The five process groups include start process group, planning process group, execution process group, supervision process group and closing process group.As shown in FIG.6-9.

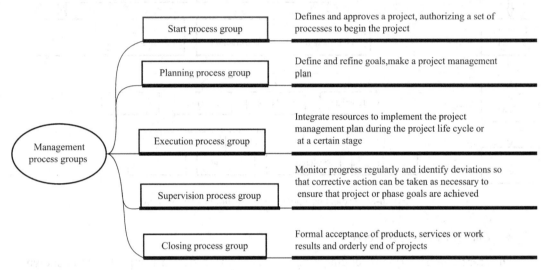

FIG.6-9　Project management process groups

- ● *Ten knowledge areas*

The ten knowledge areas include integration management, risk management, cost management, communication management, stakeholder management, scope management, procurement management, quality management, schedule management and human resource management. As shown in FIG.6-10.

6.2.3 Management functions of project participants

In the same engineering project, project management involves many stakeholders such as construction units, contractors, consulting units, suppliers, government departments, financial institutions and the public. Different participants undertake different tasks and management responsibilities at different stages. Therefore, project management is divided into roles and levels, mainly including: owner project management, supervision consultant project management, contractor project management, government project management, etc.

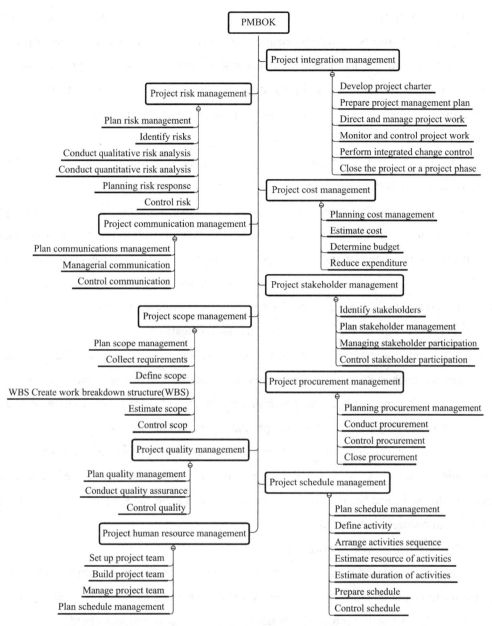

FIG.6-10 Ten knowledge areas of *Project Management the Body of Knowledge*

- ***Owner project management***

Project legal person or client to supervise and manage the whole process of project construction. The organization and management of the project legal person and the manager of the project constitute the project management of the owner.

- ***Supervision consultant project management***

The supervisor is an independent, intellectually intensive economic entity established by law. It accepts the agency of the owner and adopts economic, technical organization, contracts and other measures to supervise, coordinate and control the project construction process and participation in various aspects, in order to ensure that the project is completed successfully according to the time, investment and quality limit.The essence of consultation is to provide normative services. The Consultants are generally not directly engaged in the construction of engineering project entities, but only provide phased or whole process consultation service.

- ***Contractor project management***

The process of planning, organizing, coordinating, and controlling that completion design, construction, or supply tasks of contractor entrusted by the owner.

(1) Project management of the general contractor: Management activities of project to plan, organize, coordinate, control, direct and supervise the general contracting project according to the requirements of the general contract.It generally involves the entire process of the project.

(2) Project management of the designer: The design unit is entrusted by the owner to undertake the design tasks of the project and manage the design project according to the work objectives defined in the contract and its responsibilities and obligations.

(3) Project management of the construction party: The construction unit shall plan, organize, command, coordinate, control and supervise the systematic management activities of the whole construction process under the condition that the project manager is in charge of the implementation of the project contract and the production plan of the enterprise.

(4) Project management of the supplier: The supplier of the project materials shall take the supply project as the object of management, the scope and responsibilities defined in the supply contract as the basis, and the overall interests of the project and the supplier's own interests as the purpose of management activities.

- ***Government project management***

Government construction departments do not participate in the production activities of construction projects, but due to the large sociality, great impact and particularity of production and management of construction projects, it is necessary for the government to hold in the main behavior of construction activities through legislation and supervision, so as to ensure project quality and safeguard social public interests. The government's supervision function should through every stage of project implementation.

6.2.4　Organization structure of project management

The organization structure pattern can be described by the organization structure diagram, which reflects the organizational relationship (instruction relationship) between the components (constituent elements) in an organizational system. In a organization diagram, the rectangular

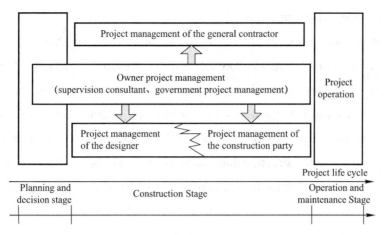

FIG.6-11　Type of construction project management

represents the work department, and the instruction relationship of the superior work department to its direct subordinate work department is represented by a one-way arrow line. The common organizational structure forms are functional, linear and matrix.

- *Functional organization structure*

In the functional organization structure, each functional department can issue work orders to its direct and indirect subordinate work departments according to its management functions. This organizational form strengthens the functional division of project management objective control and gives full play to the professional management role of functional institutions. But it is easy to contradictory instructions. Therefore, it is seldom used in project management. As shown in FIG.6-12.

- *Linear organization structure*

In a linear organizational structure, the functions of a project management organization are arranged in a straight line, with any subordinate receiving instructions from a single superior. The linear organizational structure is simple, the subordinate relationship is clear, the power is centralized, the command is unified, the responsibility is clear, the decision is quick, but the comprehensive quality of the project manager is required to be high. It is suitable for small and medium-sized projects. As shown in FIG.6-13.

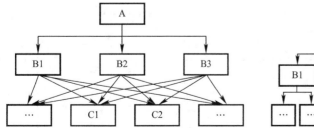

FIG.6-12　Functional organization structure

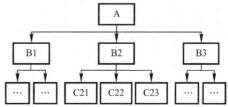

FIG.6-13　Linear organization structure

- *Matrix organization structure*

In the matrix organization structure, the instruction source of each work has vertical and horizontal. In the event of a contradiction between the direction of the vertical and horizontal working departments, coordination or decision-making is carried out by the supreme commander (department) of the organizational system. Furthermore, in order to avoid the impact of the contradiction between vertical and horizontal work department instructions on work, we can adopt the matrix organization structure dominated by vertical or horizontal work department instructions. It is mainly suitable for large complex or multiple simultaneous projects. As shown in FIG. 6-14.The matrix organizational structure example of construction enterprise as shown in FIG.6-15.

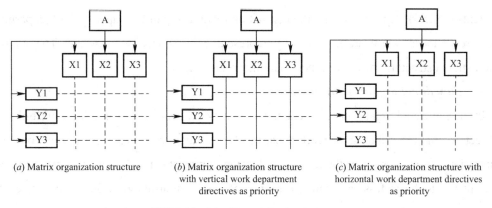

(*a*) Matrix organization structure   (*b*) Matrix organization structure with vertical work department directives as priority   (*c*) Matrix organization structure with horizontal work department directives as priority

FIG.6-14   Matrix organization structure

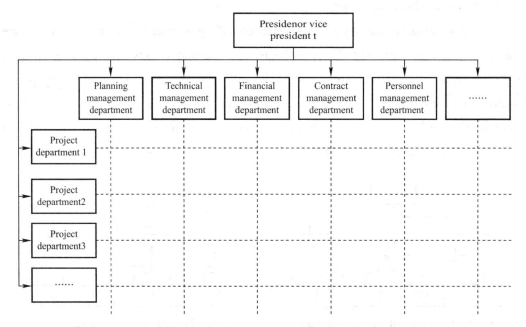

FIG.6-15   The matrix organizational structure example of construction enterprise

6.2.5　Project Management System

● *Target system*

The core contents of project management include: quality control, schedule control, cost control, change control, contract management, safety management, information management, organization and coordination, which is summarized as "four control, three management and one coordination", as shown in FIG.6-16.Quality control, schedule control and cost control constitute the project target system, as shown in FIG.6-17.

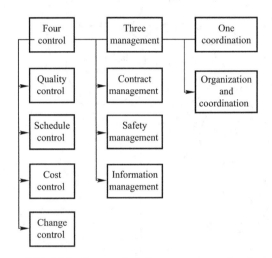

FIG.6-16　Four control, three management and one coordination

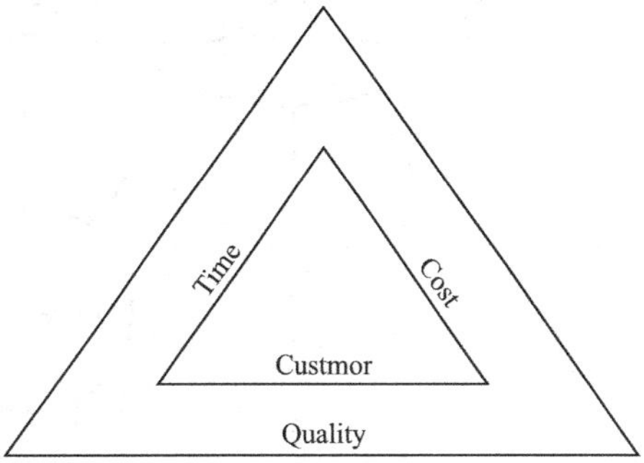

FIG.6-17　Target system of project

● *Composition of project management system*

In order to achieve the three major targets of the project, we must plan and establish a system of project management from an overall perspective. The composition of project management system as shown in FIG.6-18.

● *Operation mechanism of project management system*

The operating mechanism of project management system mainly includes motivation mechanism, constraint mechanism, feedback mechanism and continuous improvement mechanism to make the management system run normally and efficiently.

(1) Motivation mechanism

The motivation mechanism of the project management system is the core of the operating mechanism of the project management system. Its composition as shown in FIG.6-19.

(2) Constraint mechanism

The constraint mechanism of project management system depends on self-restraint ability and external monitoring effectiveness.Details as shown in FIG.6-20.

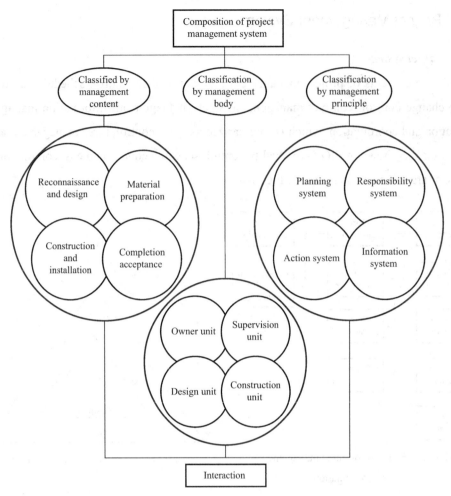

FIG.6-18　Composition of project management system

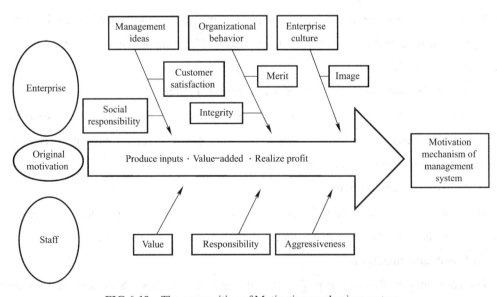

FIG.6-19　The composition of Motivation mechanism system

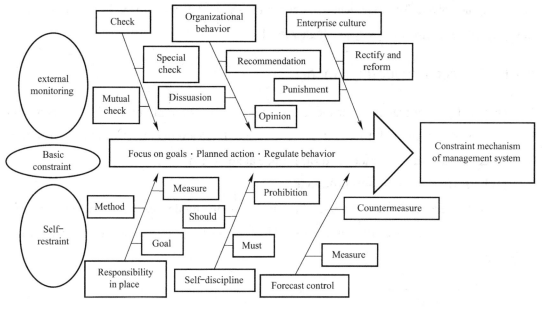

FIG.6-20   Constraint mechanism of management system

(3) Feedback mechanism

The project management system must have the ability to provide project information timely and accurately to provide the basis for managers to make decisions. Details as shown in FIG.6-21.

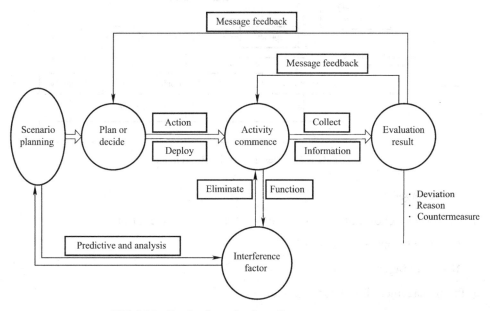

FIG.6-21   Feedback mechanism of management system

(4) Continuous improvement mechanism

With the continuous operation of the project management cycle, problems are constantly generated and solved. The experience of each stage should be summarized in time to push the next project management activity to a higher level.

6.2.6 Project management modes

Different project management modes are suitable for different engineering projects, and have certain advantages and disadvantages. At present, there are mainly DBB mode, DB mode, EPC mode, PMC mode, CM mode and BOT mode at home and abroad.

(1) DBB (Design—Bid—Build) mode

DBB (Design—Bid—Build) mode: The owner roughly divides the project into the pre-project phase and the project construction phase. The relationship between the parties as shown in FIG.6 22.

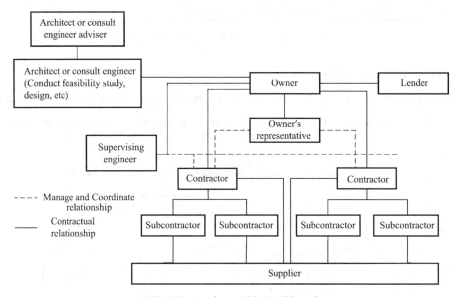

FIG.6-22 Design—Bid—Build mode

- **Advantages**

① The owner's choice of consultant and contractor is more controllable;

② Reducing the probability of risk.

- **Disadvantages**

① Pre-management is too complicated;

② Cost is higher than others modes.

(2) DB (Design—Build) mode

At the beginning of the project establishment, the owner shall, through bidding and tendering, determine the qualified contractor to contract the project according to the construction principles and requirements, and sign the contract. In accordance with the contract requirements, the contractor shall be fully responsible for the design, construction and management of the whole project. The contractual relationship of this pattern as shown in FIG.6-23.

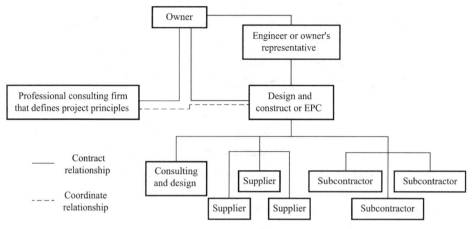

FIG.6-23　Design—Build mode

- **Advantages**

① The responsibility is in the concrete to improves the management efficiency of the project.

② Effectively reduce the total cost of the project.

③ The construction task are divided in detail to improve work efficiency.

- **Disadvantages**

① The quality of construction is greatly affected by the level of the general contractor;

② The owner lacks control over the project;

③ The construction cycle is longer and the cost is high.

(3) EPC (Engineering—Procurement—Construction) mode

EPC (Engineering—Procurement—Construction), the owner selects an engineering project company and signs a contract with it. The engineering project company contracts the whole process of reconnaissance, design, procurement and construction of the project in accordance with the contract with the owner. As shown in FIG.6-24.

- **Advantages**

① The general contractor has greater freedom of work.

② The responsibilities is clear.

③ The total contract price is fixed.

- **Disadvantages**

① The participation of the owner in the specific construction management process of the project is relatively low.

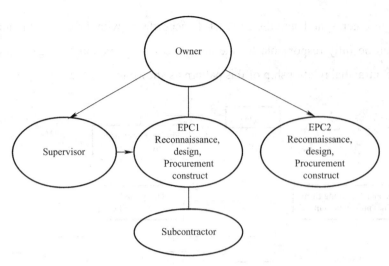

FIG.6-24 Engineering—Procurement—Construction mode

② Increased risk for general contractors.

(4) PMC (Project—Management—Contract) mode

PMC (Project—Management—Contract) mode (项目管理承包模式), the owner hands over the project to the project management contractor, who carries out the whole process management of the project from plan, project approval, bidding to construction, design and procurement, but the contractor does not participate in the detail work of the project.

- **Advantages**

① The project management contractor is experienced and has a high level of management, which helps the owner to achieve the project goals.

② PMC model can effectively improve the project organization and management structure.

③ PMC mode can effectively reduce costs and save investment.

- **Disadvantages**

The owner's ability to participate in the actual project decreases, which increases the owner's risk.

(5) CM (Construction Management Approach) mode

CM (Construction Management Approach) mode, construction while designing, also known as fast track mode. Breaking the previous continuous construction production mode. The comparison between continuous construction contracting and phase contracting as shown in FIG.6-25.

- **Advantages**

① Effectively shorten construction period.

② Communication efficiency has been significantly improved.

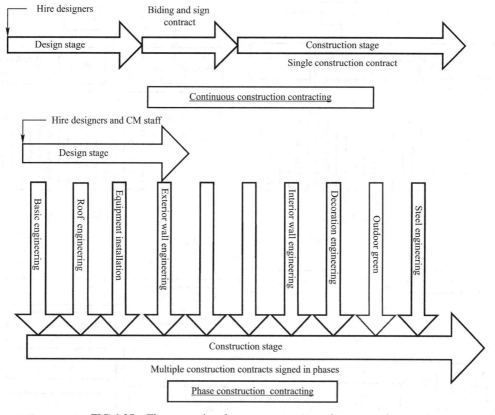

FIG.6-25   The comparison between two construction contracting

③ The employer and the contractor choose the subcontractor together, which has certain foresight and wisdom.

④ Construction is links to design can make the project more reasonable use of advanced construction technology to ensure the quality of the project.

- **Disadvantages**

① The requirement of project managers or institution is higher.

② The owner is taking a high risk.

CM mode can be implemented in two ways: Agency CM mode and At Risk CM mode. As shown in FIG.6-26.

(6) BOT mode (Building—Operation—Treasfer)

BOT mode is a kind of PPP mode, which means that the state opens the project to the outside world and gives the right of construction to a project company or investor. Then the project company or investor is responsible for the financing, organization, construction, production and operation of the project, and finally hands the project to the initiator. Its typical structural framework as shown in FIG.6-27.

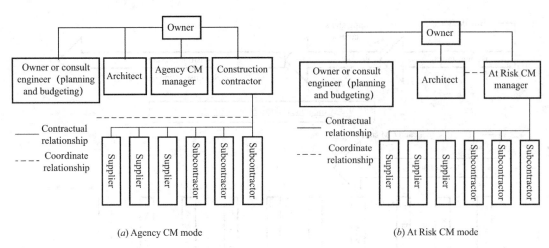

(a) Agency CM mode   (b) At Risk CM mode

FIG.6-26   Two ways to CM mode

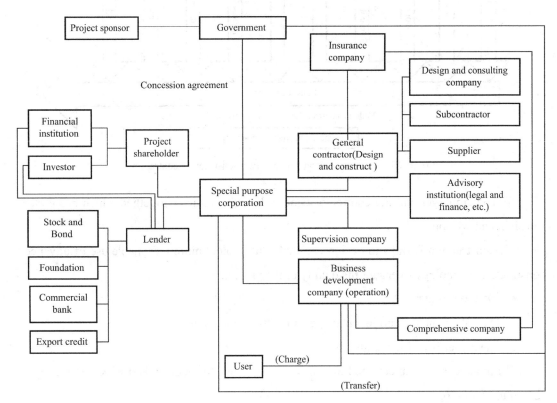

FIG.6-27   Typical structural framework of Building—Operation—Treasfer mode

● **Advantages**

① It is possible to get more financial and policy support, thus broadening the financing channels.

② It reduces the burden and risk of government debt.

③ Attract foreign funds injection and remedy the lack of domestic situation.

④ Project construction and management techniques are improved.

● **Disadvantages**

① Government control over engineering projects has declined.

② The project financing cost is high and the relationship between participants is complex.

6.3 Construction Organization

6.3.1 Introduction to construction organization

The construction of an engineering is a complex system engineering with multiple types of work and multiple specialties. In order for the construction process to proceed smoothly and achieve the expected goals, it is necessary to make good preparation for construction and carry out construction management with scientific methods.

Concept: Aiming at the complexity and diversity of the construction of the engineering, *the construction organization* is to make overall arrangements and systematic management of various problems encountered in the construction, to comprehensively deploy various activities during the construction process, and to compile the technical and economic documents that have the function of planning and guiding construction, which is the design of the construction organization.

● *Preparation works before construction*

The construction preparation work is the preparation work must be done before the construction, in order to ensure the smooth commencement of the engineering and the normal progress of construction activities. It is an important link in the construction process, not only before the start of construction, but also throughout the entire construction process.

Content: The preparation of general engineering can be summarized into six parts: investigation and research collection data, technical data preparation, construction site preparation, material preparation, construction personnel preparation, seasonal construction preparation, as shown in FIG.6-28.

6.3.2 Construction organization design

Concept: Construction organization design is a comprehensive document with technical, economic and organizational guidance for the activities of the whole process of civil engineering construction.

Task: Its task is to make a comprehensive and reasonable plan and arrangement in terms of

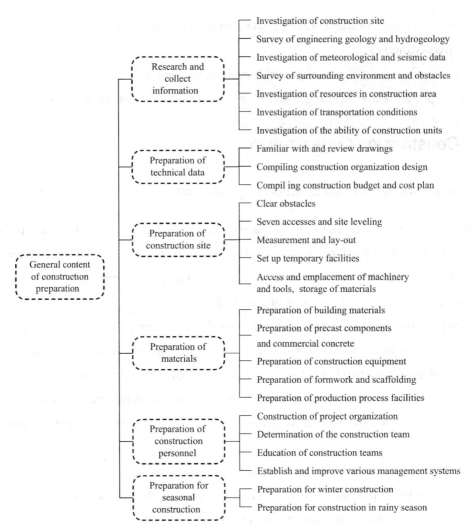

FIG.6-28 General content of construction preparations

manpower and material resources, time and space, technology and organization for the specific construction preparation and the whole construction process of the proposed engineering. Finally, scientific management is achieved to achieve the goals of improving engineering quality, speeding up engineering progress, reducing engineering costs, and preventing safety accidents.

Compilation basis: For different construction organization design, the basis is also different. The controlled construction organization design is mainly based on the content of policy and law. The implementation of the construction organization design is mainly based on specific information and relevant provisions.

It mainly includes the following aspects:

① Laws, regulations and documents related to engineering construction;

② Relevant national standards and technical and economic indicators in force;

③ The approval document of the administrative department in the area where the engineering is located, and requirements of the construction unit for construction;

④ Construction contract or tender documents;

⑤ Engineering design documents;

⑥ Site conditions, engineering geology, hydrogeology, meteorology and other natural conditions within the scope of engineering construction;

⑦ Availability of resources related to the engineering;

⑧ Production capacity, machinery and equipment, technical level, etc. of the construction unit.

Compilation principles: The compilation of the construction organization design requires mastering the principles as shown in FIG.6-29.

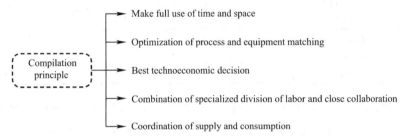

FIG. 6-29　Compilation principles of construction organization design

The basic content: According to the scale and characteristics of the proposed engineering, the complexity of construction organization design varies. However, to complete the task of organization construction, no matter what kind of construction organization design generally has the following contents:

① Engineering overview;

② Construction deployment or construction plan;

③ Construction schedule planning;

④ Plans for construction preparations;

⑤ Various resource allocation plans;

⑥ Layout plan of construction site;

⑦ Technical organization guarantee measures for quality, safety and economy, etc. ;

⑧ Major construction management plans;

⑨ Major technical and economic indicators.

Due to the different objects of construction organization design, the scope that all respects content includes above is different also. According to the actual situation of the proposed engineering, it can be changed.

Classification: According to different standards, there are different classifications of construction organization design.

- Construction organization design can be divided into two categories according to different stages: one is the construction organization design compiled before bidding (referred to as pre-bid design); and the other is the construction organization design compiled after signing the engineering contract (referred to as post-bid design).

The differences between the two types of construction organization design are shown in Table 6-3.

Table 6-3   Differences between pre-bid and post-bid construction organization design

category	Service scope	Compilation time	Compiler	Main feature	Main goals pursued
Pre-bid design	Bidding and signing	Before bidding	Operating management	Planning	Bid winning and economic benefits
Post-bid design	Construction preparation to acceptance	After signing, before starting work	Engineering management	Guiding	Efficiency and benefit of construction

- According to the compilation object, it can be divided into three categories: *general design of construction organization, construction organization design of unit engineering, construction organization design of branch engineering (sub-engineering).*

- **General design of construction organization**

Object: The general design of construction organization is based on the whole construction engineering.

Content: It is the technical and economic document used to guide the whole construction process of the engineering. It is the overall planning of the whole construction engineering, involving a wide range of content comparison overview. Generally, after the preliminary design or expansion of the preliminary design is approved.

Compiler: The general contractor shall be responsible for the general design of the construction organization, which shall be compiled jointly with the construction unit, design unit and construction subcontractor.

Compilation Procedure: The compilation procedure is shown in FIG.6-30.

- **Construction organization design of unit engineering**

Object: The construction organization design of a unit engineering is based on a unit engineering or an uncomplicated single engineering (such as a factory building, structure, or dormitory).

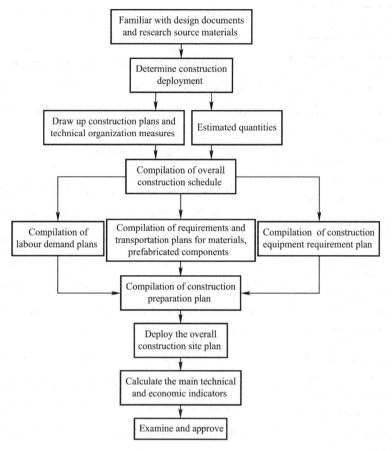

FIG. 6-30   The Compilation Procedure of the General
Design of Construction Organization

Content: The content is more specific and detailed.

Compiler: After the completion of the construction drawing design and before the commencement of the proposed construction engineering, the construction organization design of the unit engineering shall be compiled by the technical person in charge of the unit engineering.

Compilation Procedure: The compilation procedure is shown in FIG.6-31.

● **Construction organization design of branch engineering (sub-engineering)**

Object: The construction organization design of a branch engineering (sub-engineering) is based on the compilation of a branch engineering (sub-engineering). Generally, for buildings or structures with large scale, complicated technology, difficult construction, or new technology or technology, after preparing the construction organization design of the unit engineering, it is often necessary to make in-depth construction design for some important branch engineering (sub-engineering), such as deep foundation works, large-scale structure installation works, etc.

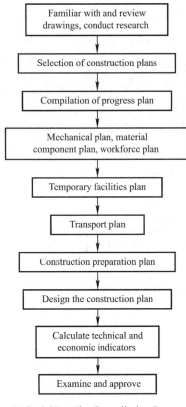

Familiar with and review drawings, conduct research

Selection of construction plans

Compilation of progress plan

Mechanical plan, material component plan, workforce plan

Temporary facilities plan

Transport plan

Construction preparation plan

Design the construction plan

Calculate technical and economic indicators

Examine and approve

FIG. 6-31 The Compilation Procedure of Construction Organization Design of Unit Engineering

Content: The content is specific, detailed, and highly operable.

Compiler: The construction organization design of the branch engineering (sub-engineering) is prepared by the technical person in charge of the unit engineering.

6.3.3 Flow construction

In the construction and installation engineering construction, there are three common construction organization methods: *sequential construction*, *parallel construction* and *flow construction*. Among them, the flow construction is a scientific, advanced and reasonable construction organization way in terms of process division, time arrangement and space arrangement.

Concept: *Flowing construction* is to decompose the entire construction process of the proposed engineering into several different construction processes, and then set up a corresponding professional work team in accordance with the construction process to adopt segmented flow operations, and the adjacent two professional teams lap parallel construction to the maximum extent.

According to the scope of flow construction organization, flow construction can be divided into:

- **Sub-engineering flow construction**

The sub-engineering flow construction is also known as detailed flow construction. It is a flow construction organized in a professional engineering, such as foundation engineering.

- **Branch engineering flow construction**

Branch engineering flow construction is also called professional flow construction. It is a flowing construction organized within a branch engineering and between various sub-engineering, such as concrete engineering, masonry engineering, and reinforcement engineering.

- **Unit engineering flow construction**

Unit engineering flow construction is also called comprehensive flow construction, which is a flow construction organized within a unit engineering and between branch engineering, such as construction and decoration engineering, and water supply and drainage engineering.

- **Group engineering flow construction**

Group engineering flow construction is also called large flow construction. It is a flow construction organized among several unit engineering.

The grading of flow construction and their mutual relationship are shown in FIG.6-32.

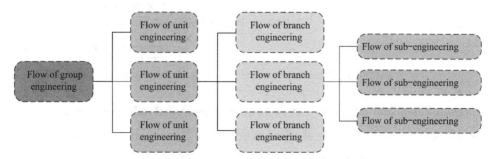

FIG.6-32   Grading diagram of flow construction

- **Flow parameters**

Flow parameters refer to the parameters used to express the status of flow construction in terms of construction process, spatial arrangement and time arrangement when organizing flow construction. It includes process parameters, spatial parameters and time parameters.

Process parameter: Classification of process parameters are shown in FIG.6-33.

Spatial parameter: Spatial parameters include working face, construction section and construction layer.

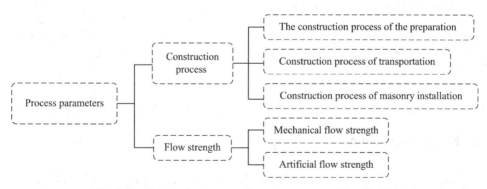

FIG.6-33   Schematic diagram of process parameter classification

Time parameters: Time parameters include flow beat, flow step distance, technical interval, organizational interval and parallel overlap time.

- **Representation of flow construction**

There are two main ways to express the flow construction, Gantt chart (horizontal channel diagram) and network diagram, as shown in FIG.6-34.

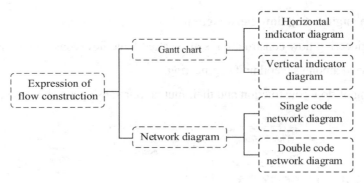

FIG.6-34　Schematic representation of flowing construction

Horizontal indicator diagram: The expression of horizontal indicator diagram of flow construction is shown in FIG.6-35. The abscissa indicates the duration of flow construction. The ordinate represents the construction process of flow construction and the name, serial number and quantity of professional task force.

Vertical indicator diagram: The expression of vertical indicator diagram of flow construction is shown in FIG.6-36. The abscissa represents the duration of flow construction. The ordinate indicates the construction section number divided by the flow construction. The oblique line segment indicates the situation of each professional task force or construction process in the flow construction.

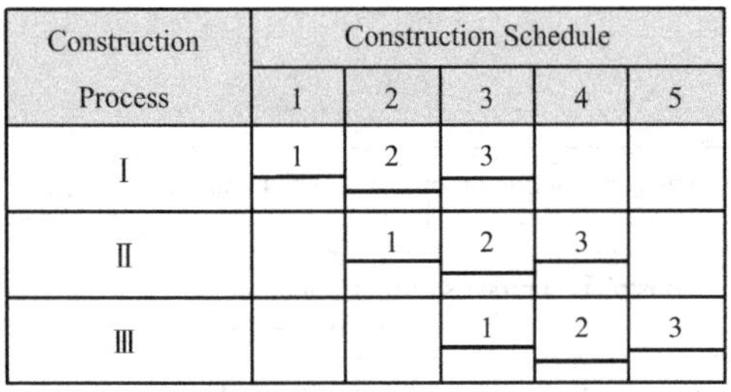

Construction Process	Construction Schedule				
	1	2	3	4	5
I	1	2	3		
II		1	2	3	
III			1	2	3

FIG.6-35　Horizontal indicator diagram

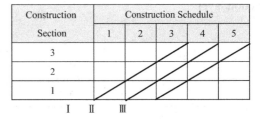

FIG.6-36　Vertical indicator diagram

6.3.4　Network planning technology

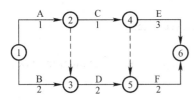

FIG.6-37　Double code network diagram

Network planning technology is a kind of work flow chart which uses the network graph to express the plan or project development sequence. It is not only a scientific planning method, but also an effective production management method. It usually has two representation methods: double code and single code, as shown in FIG.6-37 and FIG.6-38.

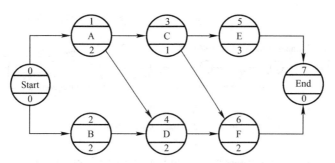

FIG.6-38　Single code network diagram

- **Double code network diagram**

The double code network diagram is composed of a number of arrows and nodes that represent work, in which each work is represented by an arrow line and two nodes, and each node is numbered. The Numbers of the two nodes before and after the arrow line represent the work represented by the arrow line.

The double code network diagram is composed of three basic elements: *work*, *node* and *line*.

Work: Work, also known as process, activity and process, can be divided into three categories:

① Work that consumes time and resources (such as building brick walls);

② Work that consumes time without consuming resources (such as curing concrete);

③ Work that consumes neither time nor resources.

The first two kinds of work are actually existing work, called "real work", which are represented by real arrow line, as shown in FIG.6-39 (*a*). The latter is an artificial dummy work, which only represents the logical relationship between adjacent work before and after. It is called "virtual work" and is indicated by a dashed arrow, as shown in FIG.6-39 (*b*).

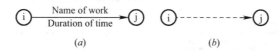

(*a*)　　　　　　　　　　(*b*)

FIG.6-39　Working schematic diagram of double code network

Node: Nodes are circles in double-coded network diagrams. It expresses the following aspects:

① The node indicates the moment when the previous work ends and the subsequent work starts, the node does not need to consume time and resources.

② The node at the beginning of the whole plan are called the start node of the network diagram, the node at the end of the whole plan are called the end node of the network diagram, and the rest are called the intermediate nodes, as shown in FIG.6-40.

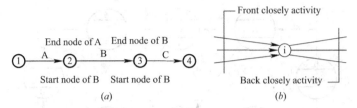

FIG.6-40   Schematic diagram of nodes

③ The number of the two nodes before and after each arrow line indicates a job. As shown in FIG.6-40, ①→② represent the work of A.

④ In the network diagram, there can be many work leading to one node, and there can also be many work starting from the same node.

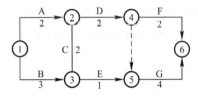

FIG.6-41   Double code network diagram

Line: In the network diagram, from the start node to the end node, it passes through a series of arrow lines and node paths in the direction of arrow line, which is called the line. In a network diagram, there are generally many lines. In FIG.6-41, there are four lines, and their total duration is shown in Table 6-4.

Among them, the line with the longest construction period is called the key line, as shown in FIG.6-41, ①→②→③→⑤→⑥ , and the work on the key line is called the key work. The total duration of the key line determines the duration of this network plan. Any delay of key work will lead to the delay of the total construction period.

Table 6-4   Total duration of each line

Line	Total duration (d)
①$\frac{A}{2}$②$\frac{B}{2}$③$\frac{E}{1}$⑤$\frac{G}{4}$⑥	9
①$\frac{A}{2}$②$\frac{D}{2}$④$_0$⑤$\frac{G}{4}$⑥	8
①$\frac{B}{3}$③$\frac{E}{1}$⑤$\frac{G}{4}$⑥	8
①$\frac{A}{2}$②$\frac{D}{2}$④$\frac{F}{2}$⑥	6

- **Single code network diagram**

The single code network diagram is to use a circle or box to represent a work. The work code, work name, and time required to complete the work are written in the circle or box. The arrows are only used to indicate the sequential relationship between the work.

The single code network diagram is composed of three basic elements: *arrow line*, *node* and *line*.

Arrow line: In the single code network plan, arrow line represents the logical relationship between adjacent works, which neither consumes time nor resources.

Node: Each node in the single code network plan represents a work, represented by a circle or rectangle. The work name, duration, and work number represented by the node should be marked in the node, as shown in FIG.6-42.

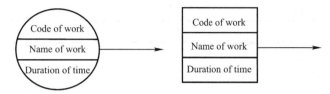

FIG.6-42　Work representation of a single code network diagram

Line: The lines of the single code network diagram have the same meaning as the lines of the double code network diagram. Similarly, the line with the longest duration is called the key line, and the work on the key line is called the key work.

Single code network diagram and double code network diagram are two kinds of complementary and characteristic expression methods. In some cases, the application of single code representation is relatively simple, and in some cases, the use of double code representation is more clear.

Questions

- Briefly describe content of every construction procedure.
- What are the organization structures of project management? What are the characteristics of each other?
- Briefly describe the content of construction preparation.
- Try to describe the concept, basic content and classification of construction organization design.
- What is flow construction?
- What are the expressions of flowing construction?
- Draw the double code network plan according to the logical relationship of each construction process in Table 6-5.
- Draw the single code network plan according to the logical relationship of each construction process in Table 6-3.

Table 6-5　working relationship table

Name of the work	Duration of time	Front closely work	Back closely work
A	2	—	B、C
B	3	A	D
C	2	A	D、E
D	1	B、C	F
E	2	C	F
F	1	D、E	—

Reference List

[1]　Shengxing Wu. Construction regulations (3rd Edition)[M]. Beijing: Higher Education Press (HEP), 2017.

[2]　Hongfeng He, Liansheng Zhang, Yu Yang. Building Regulations Tutoria (4th Edition)[M]. Beijing: China Construction Industry Press, 2018.

[3]　Wei Xu, Jiayun Wu, Jianwen Zou. Civil Engineering Project Management[M]. Shanghai: Tongji University Press, 2010.

[4]　Shizhao Ding, Engineering Project Management (2nd Edition)[M]. Beijing: China Construction Industry Press, 2014.

[5]　Chongqing University, Tongji University, Harbin Institute of Technology. Civil Engineering Construction[M]. (3rd Edition) Beijing: China Construction Industry Press, 2016.

[6]　Han Yuwen, Li Wang, Li Junfeng. Civil Engineering Construction Organization [M]. Beijing: China Building Materials Industry Press, 2017.

[7]　Hu Changming, Li Yalan. Building Construction Organization[M]. Beijing: Metallurgical Industry Press, 2016.

[8]　Zhang Huamin, Ji Fanrong, Yang Zhengkai. Building Construction Organization [M]. (3rd Edition) Beijing: China Electric Power Press, 2018.

CHAPTER 7
HAZARD MITIGATION OF CIVIL ENGINEERING

Earthquake, fire, storm, flood, debris flow, explosions and other disasters are common natural disasters in the field of civil engineering. In this chapter, these engineering disasters are introduced and analyzed, and some measures for disaster prevention and reduction are put forward.

7.1 Introduction of Hazard

7.1.1 The definition of disaster

Disaster refers to all kinds of phenomena that cause damage to the survival and social development of human beings because of natural, man-made, or integrated human and natural reasons. Disaster is an extreme form of expression of the movement, change and development of things, which are characterized by damaging the interests of human beings and threatening the survival and sustainable development of human beings.

7.1.2 The type of disaster

There are various kinds of disasters and different classification methods. But from the perspective of the mechanism of disaster formation, it can be classified into two categories, natural disasters and man-made disasters.

Natural disasters are the result of physical movement and change in nature. It is called the disaster, because these are the result of natural phenomena beyond a certain limit, and caused catastrophic damage to the survival of human beings and the environment.

Man-made disasters refer to disasters caused mainly by human factors, they can be divided into individual behavior disasters and social behavior disasters.

(1) The main natural disasters include the following types:

1) Geological disasters: earthquakes, volcanic eruptions, landslides, mudslides, ground sub-

sidence, etc.

2) Meteorological disasters: rainstorm, floods, hail, tornadoes, droughts, snowstorms, frost, etc.

3) Bio-disasters: forest fires, sandstorms, pests, acute infectious diseases, etc.

4) Astronomical disasters: celestial body collision, abnormal solar activity, etc.

FIG.7-1   Wenchuan earthquake site   **FIG.7-2**   Snowstorm in southern China in 2008

(2) The main man-made disasters include the following types:

1) Ecological and environmental disasters: air pollution, greenhouse effect, water pollution, soil erosion, population explosion, etc.

2) Engineering accidents: explosion, nuclear leakage, house collapse, traffic accident, etc.

3) Political and social disasters: war, social violence and turmoil, financial storm, etc.

FIG.7-3   Haze in Beijing   **FIG.7-4**   Aerial photography of the Chernobyl nuclear
power plant

7.1.3   The harm of disaster

Disaster is the great enemy of human existence, production and civilization construction, they are often mercilessly attacking human beings with their great energy. After the birth of human, the disaster has always been accompanied, in the vast historical documents of various regions and countries of the world, countless tragic records of disaster have been left. The harm of disasters can be attributed to the following three categories:

Casualties caused: Natural disasters directly harm human life and health. A serious disaster may cause millions and billions people to be hit and cause huge casualties. For example, in January 1556 Shaanxi Huaxian, Tongguan earthquake caused 830 thousand deaths; In July 1931, the catastrophic flood in the Yangtze Huaihe River Valley caused 220 thousand deaths in one month.

Huge economic losses: In recent decades, due to human intervention, the frequency of disasters has increased. On the other hand, as the population is concentrated in the city, the accumulation of wealth has increased rapidly, and the economic losses caused by the disaster have increased unabated. Damage to infrastructure such as houses, factories, railways, highways, electric power and communications will cause huge indirect economic losses besides huge direct economic losses. It is mainly due to the economic loss caused by shutdown and stop production, and sometimes more than the direct economic loss.

Destruction of environmental resources: Disasters and environment are closely interacted, deterioration of environment can cause natural disasters, natural disasters will also promote environmental deterioration, thereby affecting the development of social economy. Such as desert migration cause land desertification, water depletion lead to land desolation, land salinization and so on. All these make the environment worse and affect the further development of production.

FIG.7-5 Loess Plateau

7.2 Various Engineering Disasters and Prevention and Control Measures

The most common disasters in the field of civil engineering include earthquakes, wind disasters, and geological disasters (debris flows, landslides, etc.). The following describes different types of engineering disasters and disaster prevention.

7.2.1 Earthquake disaster and earthquake resistance

Earthquake is a natural phenomenon. It is called the earthquake due to the sudden rupture of rock in some underground area, or the vibration caused by partial rock collapse, volcanic eruption and so on.

According to statistics, there are about more than 5 million earthquakes on the earth every year, namely the earthquake to happen thousands of times a day. Most of them are too small or too far away to be felt; there are about ten or twenty earthquakes that can cause serious harm to human beings; about one or two earthquakes that can cause particularly serious disasters. Table 7-1 shows the catastrophic earthquake disaster since the 20th century.

Table 7-1 Catastrophic earthquakes since 20th Century

Time and place of occurrence	Magnitude and secondary disaster	Cause loss
2004 Indonesian Sumatra	Level 8.7, and cause a tsunami	The number of deaths exceeded 150 thousand
Taiwan, China, 1999	Level 7.6	2400 people died and 11300 injured
Turkey, 1999	Level 7.4	Death of 13 thousand people
Tangelos, 1994	Level 6.7	55 people died and more than 7000 injured
Armenia, former Soviet Union, 1988	Level 6.9	55 thousand people died and 13 thousand seriously injured
Tangshan, China, 1976	Level 7.8	Death of 242 thousand people, more than 16.4 million people
1970 Peru Chimbote	Level 7.8, causing the most violent debris flow in history	Death of 23 thousand people
Chinese Haiyuan, 1920	Level 8.5, causing mass landslides	Death of 234 thousand people
Italy Messina, 1906	Level 7.5, triggering a tsunami	Death of 83 thousand people

Most earthquake-related property damage and deaths are caused by the failure and collapse of structures due to ground shaking (FIG.7-6). The level of damage depends upon the extent and duration of the shaking. Other damaging earthquake effects include landslides, the down-slope movement of soil and rock (in mountain regions and along hillsides), and liquefaction.

(1) Seismic knowledge

Hypocenter focus: An earthquake in the earth's interior.

Epicenter: The projection of the epicenter on the surface of the earth, or where the ground is facing the source.

Isoseismal: A line on a map connecting points at which earthquake shocks are of equal intensity.

FIG.7-6   The collapse of structures

Epicentral distance: The straight line distance from any point on the ground to the epicenter is called the epicentral distance. Earthquakes of the same size, the smaller the epicentral distance, the heavier the impact or damage.

Magnitude: Seismic magnitude is an indicator of the magnitude or strength of the earthquake. It is the measure of how much energy is released by the earthquake, and it is one of the basic parameters of the earthquake.

Intensity: Seismic intensity refers to the intensity of the ground and all kinds of buildings affected by an earthquake in a certain area.

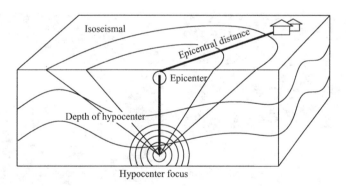

FIG.7-7   Seismic tectonics

* Supplementary instructions: For an earthquake, there is only one magnitude to indicate the magnitude of the earthquake, but its influence on different locations is different and the intensity of the earthquake varies.

(2) The damage of earthquake to civil engineering

China, located at two most active seismic belts, facing the Circum Pacific seismic belt in the east and being on the path of Eurasia seismic belt in the west and southwest, is one of the most earthquake-prone countries in the world. The cities located in strong earthquake regions constitute a substantial proportion.

1) The damage to civil buildings

- **Masonry building:** For brick masonry walls the cracks were liable to occur due to the lower tensile and shear strength and the poor ductility. In case of slight destruction, minor cracks occurred near the corners of door/window openings; in case of rather severe destruction, diagonal shear cracks or X-shaped cracks occurred on the longitudinal and transversal walls, on the walls between and under windows, or on the whole walls; in case of very severe destruction, the walls were destroyed and collapsed. For buildings with inner frames, due to the inharmonious deformation between infill walls and the frames, cracks were liable to occur on the walls, mainly the horizontal cracks under the frame beams and diagonal cracks on the walls. The tie-columns and ring beams played an effective and integral constraint role, though some cracks occurred at the connection ends and concrete were crushed. The penthouses were damaged in varying degrees or collapsed due to whipping effect. The walls in the staircases were liable to be damaged due to the openings for electricity meters. The walls were partially destroyed and the decorated materials fell off due to the pounding effect of buildings on both sides of the expansion joint which was not wide enough.

FIG.7-8　The damage to masonry structures

- **Reinforced concrete structure house:** The reinforced concrete structures consisted of the structures with frame, shear wall, frame-shear wall, bottom frame, special-shaped column frame and inner frame, etc. For the shear wall structure, the damage included the shear failure of the spandrel and the damage to the lower part of shear wall. For the frame structure, most damage occurred at the column ends and beam-column joints (presenting strong beam weak column) and the main forms of earthquake damage were:

horizontal cracks on column ends, diagonal or cross cracks on column ends and joints, even rupture, dislocation, spalling of concrete and buckling of bars; destruction in the form of short columns due to irrational laying of infill wall or high-positioned windows; slight damage to the frame beam, and longitudinal bending cracks or diagonal shear cracks at beam ends. For the frame infill wall structure, the main damage forms were: diagonal or cross cracks on the end walls and piers or near the corner of the door openings, or horizontal cracks above or below the window openings, collapse of the walls with the large areas and big window openings and destruction of arch walls.

The damage to the stairs in frame structures was a new problem. In the previous practice, only the static analysis was made to the stairs without considering the seismic calculation. But in reality, the stairs added the lateral stiffness to the frame structure and the stair slab suffered the repeat tension and compression action under the horizontal earthquake action. In case of slight damage, one or two horizontal cracks occurred on the stair slab and shear cracks on stair landing beam-slab; in case of rather sever damage, the bars in the stair slab were buckled, even ruptured, the concrete of stair landing beam- slab fell and bars were exposed; in case of very severe damage, the stair slab was pulled away. The projecting part from the top of the frame, especially the slender towers, was found to suffer a more serious damage in the quake.

FIG.7-9　The damage to reinforced concrete structures

- **Large span space structure:** Most stadiums/gymnasiums are space frame structures (spaces truss or lattice shell). Featured by light-weight roof and superior integrity, the space frame structure demonstrated its superior earthquake resistance capacity. In Wenchuan earthquake some stadiums/gymnasiums kept intact in the main and were used as the shelters for earthquake victims.

FIG.7-10　People take refuge in the gym

2) The damage to industrial buildings

In the quake-hit area, single-storey factory buildings, including both concrete and steel structures. For bent column, the RC bent columns (rectangular, I-shaped, open-webbed or lattice type) and I-shaped steel bent column were used; for roof truss, the PC truss, herringbone steel truss, PC thin webbed beam and steel truss were used; and for roof cover, the precast ribbed or channel roof slab and lightweight colored sandwich steel plate or corrugated steel sheet were used. The main damage forms of single factory buildings were: falling-off of precast ribbed roof slabs, partial or whole collapse of roof structures; buckling of column braces; buckling or break-ing of transverse braces at the bottom chord; and serious collapse of exterior nonstructural wall (mainly the brick masonry of the gable).

The damage to factory buildings of Dongfang Steam Turbine Works in Mianzhu was the representative. Most of them using traditional bent columns, PC truss. The precast ribbed roof slab collapsed entirely, causing serious losses of lives and facilities. Fortunately, seven similar buildings survived the earthquake thanks to the lightweight colored sandwich steel plate or cor-rugated steel sheet replacing the precast ribbed roof slab a few years ago. The factory buildings with steel truss remained basically well thanks to the satisfactory bracing system in spite of using the heavy precast ribbed roof slab. The factory buildings with portal frame withstood the earth-quake and remained basically well.

3) The damage to Municipal projects

Bridge engineering is one of the lifeline projects, and the damage of lifeline engineering (generally refers to the destruction of urban water supply, power supply, gas supply, telecommu-nications, transportation and other infrastructure) causes great difficulties for disaster relief work after the earthquake, which aggravates the secondary disasters. Especially for the modern city, it will affect the operation of its production and lead to huge economic losses.

FIG.7-11   The damage to industrial buildings

San Fernando earthquake occurred on the morning of February 9, 1971 a magnitude of M6.7. Although the earthquake magnitude is not high, is located in the meizoseismal area the ground motion large ground deformation and strong, resulting in high-rise structure, bridge collapse and lifeline engineering damage. The biggest lesson in bridge earthquake damage is the serious collapse of two interchange projects, one is Golden State's high speed road 210 interchange hub; the other one is Golden State's high speed main road and interstate Expressway No. 14 interchange hub, mainly due to the obvious lack of transverse stirrups.

The main factors leading to the bridge damage in Tangshan earthquake in 1976 are pier failure, bearing failure, beam collision and large relative displacement of adjacent pier.

FIG.7-12   Destruction to municipal engineering

(3) Earthquake resistance

In terms of seismic resistance of the civil engineering, the seismic design and inspection standards of the project were improved by revising the seismic fortification intensity of each region and proposing the seismic fortification target clearly. The research and application of structural vibration control in the past decade has become a hot spot in the field of seismic engineering. The traditional seismic design method of the structure relies on increasing the strength and deformation capacity of the structure to resist earthquakes. The damping control method which is safe, reliable, more effective, economical uses seismic isolation, energy consumption, external force application, structural dynamic characteristics adjustment, etc. to reduce structural seismic response, it has becoming an active and effective technique for civil engineering disaster prevention and mitigation.

Seismic isolation and energy dissipation are the most mature and widely used methods in civil engineering. Seismic isolation is the separation of seismic forces from the structure by some kind of isolation device to reduce structural vibration. Seismic isolation devices are widely used in the foundation of buildings and bridge abutments, and are rich in forms such as rubber cushion seismic isolation device, ball seismic isolation device, powder pad seismic isolation device, and steel seismic isolation device. The energy dissipation-seismic reduction of structures is to reduce the vibration of the structure by using certain energy-consuming devices or additional substructures to absorb or consume the energy transmitted by the earthquake to the main structure. The setting of the energy-consuming devices or the additional substructure won't change the main structural system, and can simultaneously reduce the horizontal and vertical seismic action of the structure, it isn't limited by the type and height of the structure and can be widely adopted in both new construction and seismic strengthening of buildings. The energy dissipation-seismic reduction system is suitable for high-rise buildings, super high-rise buildings and high-rise structures. Energy-consuming devices include various energy-consuming braces, energy-consuming shear walls, friction dampers, viscoelastic dampers, and viscous dampers.

FIG.7-13　Rubber cushion seismic isolation device

7.2.2　Wind disaster and resistance

The wind is that the air in a place where the pressure is high and flows to a place where the pressure is low. The types of winds commonly found in nature are tropical cyclones, monsoons and tornadoes. In addition to sometimes causing a small number of casualties and disappearances, wind disasters mainly destroy houses, vehicles and boats, trees, crops, communications and electrical facilities.

(1) Beaufort wind scale

The Beaufort wind scale is an international scale of wind velocities ranging for practical purposes from 0 (calm) to 12 (hurricane force). The average wind speed reaches 6 or above (wind speed exceeds 10.8 m/s), and the instantaneous wind speed reaches 8 or higher (wind speed is greater than 17.8 m/s), the wind that has a serious impact on life and production is called gale.

Table 7-2　Wind scale and wind speed list

Wind scale	Name	Wind speed	
		km/h	m/s
0	Calm	<1	0~0.2
1	Light air	1~5	0.3~1.5
2	Light breeze	6~11	1.6~3.3
3	Gentle breeze	12~19	3.4~5.4
4	Moderate breeze	20~28	5.5~7.9
5	Fresh breeze	29~38	8.0~10.7
6	Strong breeze	39~49	10.8~13.8
7	Moderate gale	50~61	13.9~17.1
8	Fresh gale	62~74	17.2~20.7
9	Strong gale	75~88	20.8~24.4
10	Whole gale	89~102	24.5~28.4
11	Storm	103~117	28.5~32.6
12	Hurricane	118~133	32.7~36.9

The level of wind disasters is generally divided into 3 levels:

● **General gale:** It is equivalent to a gale of 6~8, which mainly destroys crops and generally does not cause damage to engineering facilities.

● **Strong gale:** It is equivalent to the gale of 9~11, in addition to destroying crops and trees, it can cause different degrees of damage to engineering facilities.

● **Enormous gale:** It is equivalent to a gale of 12 or more, in addition to destroying crops and trees, it can cause serious damage to engineering facilities, ships and vehicles, and seriously threaten people's lives.

(2) The damage of wind to buildings and structures

To the various natural disasters in history, the wind seems not easy to arouse people's fear, in fact, its harm is not at all under the flood and the earthquake. According to the World Meteorological Organization (WMO) report that the number of deaths is 20 thousand ~3 million people from typhoons in the world a year. The average economic loss of the Western Pacific coastal countries is about 4 billion dollars a year due to typhoons. The destructive effect of wind on buildings and structures is mainly reflected in the following aspects:

1) The damage to building construction

① The destructive effects on tall buildings' structures. Strong winds may deform the tall buildings, causing severe damage to the enclosure and/or severe shaking.

② The destruction of exterior wall finishes, door, window and glass curtain wall.

③ The destruction of high-rise structures such as masts and TV towers. This is mainly because: the stiffness of the mast structure is small, and it is easy to generate a large amplitude of vibration under wind disaster.

④ The damage to small houses, especially those with light roofs.

2) The damage to lifeline projects

The wind may destroy urban municipal facilities, communication facilities and transportation facilities, resulting in power outage, water break and traffic interruption. Bridges are also easily destroyed in high gales, the recorded wind damage to bridges dates back to 1818, and the Dryburgh Abbey Bridge in Scotland was destroyed by the wind. Between 1818 and 1940, there were many accidents in the world where bridges were destroyed by wind.

FIG.7-14 Destruction of houses

FIG.7-15 Destruction of Tacoma Narrows Bridge by wind

3) The damage to ancillary buildings such as billboards and placards

Billboards and placards are often built on the top of the main building. They are often vertical cantilever structures. The wind receiving area of them is relatively large, and the bending resistance of the roots is often insufficient so that they are easy to fall over when the wind is strong.

(3) Wind disaster resistance

The typhoon is one of the main disasters of wind disasters, at present, human beings still have no ability to eliminate it. No matter from the intensity, or the area, typhoon is the most destructive weather systems in the atmosphere of, and the typhoon has a strong seasonal, regional and sudden impact on China. Although typhoon disaster is extremely complex, the development of modern science and technology, especially the emergence of meteorological satellites, enables people to closely monitor

FIG.7-16 Destruction of billboards

and track typhoons and make timely forecasts to avoid some possible damage caused by windstorms. For structures such as offshore platforms, sea-crossing bridges, and docks, typhoons must be considered in the design. Perfecting water conservancy projects can avoid secondary disasters and derivative disasters caused by typhoons, interruption of typhoon disaster chain is very obvious to reduce typhoon losses.

For flexible structures with large windward area such as high-rise buildings, large span structures, flexible large-span bridges, transmission towers and aqueducts, wind-resistant design and seismic design are equally important.

7.2.3　Geological disaster and resistance

Geological disasters are the hazards to human survival and development due to geological processes. Natural variability and artificial effects can lead to changes in the geological environment or geological body. When this change reaches a certain level, the consequences will cause harm to humans and society.

(1) Type of geological disasters

There are many types of geological disasters, and the classification methods are very complicated.

According to the cause of the disaster they can be divided into:

① Geological disasters mainly caused by natural variation, such as volcanic eruptions, earthquakes, ground collapse, etc.

② Geological disasters mainly caused by human actions, such as landslides, ground collapse, etc. caused by road repairing slopes, construction of water conservancy facilities, excessive exploitation of groundwater, mining, etc.

③ Geological disasters caused by changes in climatic conditions, such as heavy rains and extreme precipitation events, will lead to increased landslides, collapses, mudslides, etc.

According to the geological environment or the speed of geological body changes, geological disasters can be divided into two types: sudden type and slow type. The former such as landslides, collapses, mudslides, earthquakes, volcanic eruptions, geotechnical accidents; the latter such as soil erosion, land desertification and swamping, soil salinization, etc., also known as environmental geological disasters.

FIG.7-17　Geological disasters

FIG.7-18　The road destroyed by mudslides

(2) Hazards of geological disasters

Geological disasters don't only threaten towns, villages, industrial and mining buildings, but also pose serious threats to civil engineering infrastructure such as transportation, hydropower stations and reservoirs. There are more than 100 large and medium-sized landslides and 1386 mudslides sites during China's railways. The annual renovation cost exceeds 100 million Yuan. Nearly a thousand hydropower stations and hundreds of reservoirs in the country have suffered from geological disasters such as collapse, landslides and mudslides. In addition, disasters such as land subsidence pose threats to more than 20 urban buildings in Shanghai, Tianjin and Beijing.

(3) Geological disaster resistance

There are three methods for prevention and control of landslides: reinforcement by rock bolt, establishment of slope protection or retaining wall, and reduction of slope. Most of the time, these three methods should be used at the same time. The reinforcement by rock bolt enhances the integrity of the soil by driving the rock bolt into the unstable soil. The rock bolt penetration depth must exceed the thickness of the unstable soil. Slope protection can protect the slope from direct rain, reduce the intensity of rain erosion, and protect the soil. Retaining walls provide lateral support for the soil and avoid gravity landslides. The reduction of slope method reduces the mass of unstable soil by excavating the mountain and reducing the slope of the mountain. On the other hand, the shearing force on the sliding surface is reduced when the slope of the sliding surface is slowed down.

Comprehensive prevention and control measures shall be taken for the prevention and control of debris flow. Measures such as crossing, guiding, blocking and soil and water conservation shall be adopted according to the specific conditions of the protected area.

FIG.7-19   The prevention and control of landslides

FIG.7-20   The prevention and control of debris flow

7.2.4   Other disasters

(1) Fire

Fire is indispensable in the production and living activities of people, the progress of human beings and the development of society can hardly do without fire. However, if the fire is out of control, it will jeopardize human beings, cause loss of life and property, and lead to conflagration.

Statistical analysis of fire damage show that the most frequently happened and the most serious loss is a building conflagration. All kinds of buildings are the places for people to produce and live, and the place where the property is very concentrated, therefore, the losses caused by building fires are very serious, which directly affects people's activities.

According to the statistics, the world's developed countries have lost billions of dollars to billions of dollars a year, accounting for 0.2%~1.0% of the gross national economy. From 1976 to 1980, there were 3 million 80

FIG.7-21   Damage to buildings by fire

thousand fires on average each year. The number of deaths was close to 9000, and the direct economic loss was immeasurable.

The destructiveness of the fire is not only the result of the death of the person and the destruction of the property, but also the cause of the serious indirect loss. Fire can cause personal injury and damage to property. It can be measured in terms of money, but it is difficult to measure with money that result in factory closures, workers' unemployment and school closures. In addition, the pollution caused by the fire to the environment and the serious destruction of the ecological balance are difficult to eliminate in the short term.

(2) Big flood

Floods are one of the most frequent hazards in the China. Flood effects can be local, impacting a neighborhood or community, or very large, affecting entire river basins and multiple states. Some floods develop slowly, sometimes over a period of days. Flash floods can develop quickly, sometimes in just a few minutes or without any visible signs of rain. The causes of the formation of flood disasters can be attributed to two aspects of nature and society. Nature factors: changes in the weather system; unevenness of rainstorm time and regional distribution; the effects of tropical storms and typhoons and the changes in topography and geomorphology, etc. The main social factors are the intensification of human activities, which include deforestation, vegetation destruction, protected farmland etc. Eventually, those activities aggravate soil erosion, reduce the storage area, block the river flood, resulting in increased peak flow.

FIG.7-22　The damage caused by the flood

The various types of flooding include riverine flooding, coastal flooding, and shallow flooding. Common impacts of flooding include damage to personal property, buildings, and infrastructure; bridge and road closures; service disruptions; and injuries or even fatalities. The distribution of mountain torrents is generally larger than the debris flow, including the harm to cities and towns, traffic and transportation, mine and farmland.

7.3　Engineering Structure Detection, Identification and Reinforcement

Some scholars have proposed the famous "five-fold effect" law, that is, investing 1 Yuan to build a new building, then the cost for reinforcement in the future is 5 Yuan. This shows the importance of engineering structure detection, identification and reinforcement. The impact of engineering structures caused by earthquakes, winds, fires, freezing, corrosion and improper construction will eventually be solved by engineering structure detection, identification and reinforcement.

7.3.1   Structural disaster detection and identification

The structural detection and identification is to test the durability of the engineering structure by various methods, identify its safety, reliability, the conclusion of its identification level and whether it needs reinforcement. The procedures are: inspection task entrustment, investigation, preparation of detection plan, on-site detection, issue of detection report, identification rating and issue of identification report.

The reconstruction and reinforcement of the existing building is based on the original building. Therefore, before the reconstruction and reinforcement, the existing building should be identified and evaluated to fully understand the structural performance and safety risks of the building (such as material strength, structural measures, cracks, deformation and the conditions of use, etc.), and make a comprehensive assessment of the reliability of the structure.

7.3.2   Performance of materials in disasters

In the post-disaster detection and reinforcement research of engineering structures, the primary concern is the changes in the properties (such as strength, elastic modulus, constitutive relationship, etc.) of the materials after the disaster. For this reason, many studies have been done at home and abroad. Some conclusions have been got qualitatively and quantitatively, but the system is still not enough, so disaster materials science has not yet formed a special discipline in the field of civil engineering. However, in the practice of reinforcement design, identification and consulting of engineering structures, this knowledge is indispensable.

Disaster materials science generally involves the general mechanical properties of civil engineering materials, such as internal cracks and failure mechanisms of concrete, internal structural failure mechanisms of steel

FIG.7-23   Detecting the strength of concrete after a fire

bars, general failure mechanisms of masonry, etc.; effects of dynamic loads on materials, such as the effect on concrete and steel caused by the fatigue and impact of steel bars ; the effect of fire on material properties, such as the effect on concrete or steel, the effect on the bond between concrete and steel; the effect of freezing on the properties of materials, such as the mechanical properties of frozen concrete; the effect of corrosion on material properties, etc.

7.3.3 Engineering structure transformation and reinforcement

Engineering structural transformation and reinforcement is a discipline that studies how to restore the damaged engineering structures or make them adapt new functions. Due to the rapid development of the construction industry and modern science technology, the development of this discipline is extremely rapid.

For the reinforcement and reconstruction of existing buildings, the safety of the building should be improved through scientific and reasonable structural reinforcement design to prevent the destruction of building structures. In addition, the safety reserve of the building structure can also be improved by selecting safe reinforcement materials. The investment in improving the durability of the building structure is economical as well as can effectively extend the service life of the building, so the structural reinforcement design is increasingly valued by the government, experts and scholars.

At present, the cases which need for reinforcement are: the decline in the use function caused by the service life, the change of use of the building, the damage caused by disasters or accidents, and the higher comfort requirements to the building.

There are many reinforcement methods for buildings, which are applicable to different engineering structures.

FIG.7-24 Transforming the industrial plant into a library

(1) The main reinforcement methods for concrete structures are:

- Expanding the section. The method of expanding the section is the easiest way. By increasing the section of the concrete beam and column, the bearing capacity of the beam and column is improved. The disadvantage of this method is that there is a certain reduction in building clearance.

- Bonding the steel or carbon fiber outside the members. By attaching a steel sheet or carbon fiber to the surface of the concrete member (generally in the tension zone), it will works in conjunction with the concrete to improve the load carrying capacity of the member, but requires special fireproof treatment.

- Wrapping the steel around the members. This reinforcement method is to wrap the steel in the four corners or two sides of the components (such as concrete columns), and the cross-sectional dimension of the components is not much increased, but the bearing capacity of the concrete columns can be greatly improved.

FIG.7-25 Bonding the carbon fiber outside the members

FIG.7-26 Wrapping the steel around the members

- Prestressed reinforcement. By prestressing the building components, it is possible to offset some of the applied loads and act as an unloading action. It is especially suitable for large span structures.

(2) The main reinforcement methods for masonry structures include the addition of wall column and steel mesh cement mortar method. In addition, we can also expand the section and wrap the angle steel to reinforce the brick columns.

FIG.7-27　Expanding the section of masonry structures

(3) The main reinforcement methods for steel structures are:

- Changing the structure calculation graph.
- Increasing the section of members.
- Strengthening the connection between the members.

The development of civil engineering relies to a large extent on the application and development of new materials and new technologies. As for the reinforcement materials, the conventional materials include steel plates, steel sections, high-strength bolts, etc. In the field of reinforcement and modification of existing structures, the materials not only need to be economy, but also easy to construct, and the structural bearing capacity after construction is required to be significantly improved. Composite materials play an increasingly important role due to their excellent mechanical properties and wide applicability. Compared with steel, the reinforcement of the structure with composite materials has the following advantages: lighter weight, higher strength, higher elastic modulus, better fatigue resistance, better corrosion resistance, better durability, thinner thickness, simpler construction, etc.

7.4　The New Achievements and Development Trend of Disaster Prevention and Reduction

In 1987, the 42nd session of the UN General Assembly adopted Resolution No. 169, which initiated an international activity named World Decade for Natural Disaster Reduction from 1990 to 2000, the last decade of the 20th century. Its purpose is to minimize the losses caused by various natural disasters in the world, especially those caused by natural disasters in developing countries, through concerted international efforts. This kind of extensive collaboration based on the purpose of joint disaster reduction requires governments, scientific and technical groups and various non-governmental organizations to respond positively to the call of the UN General Assembly and extensively carry out various forms of international cooperation under the unified leadership and coordination of the UN, through technology transfer and assistance, project demonstration, education training and carry out various other activities to mitigate natural disasters, thereby enhancing each the disaster prevention and reduction of the country, especially the third world countries.

Since the beginning of the 20th century, city as the most complicated social system, has begun to be affected by more and more external disturbances, including natural disasters such as floods and earthquakes, as well as events such as terrorism and global social crisis, all of them pose a great threat to the normal operation of the urban system. Under this background, scholars have proposed the concept of resilient city, in order to propose

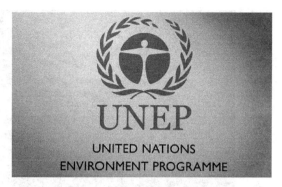

FIG.7-28   The logo of United Nations Environment Programme

key measures for the solution of urban problems. Holling (1973) first presented the idea of resilience as the ability of a system to deal with severe changes and remain to function. The so-called "resilience" is the resistance to impact and restoring ability of the system itself. The resilient city described by scholars at home and abroad mainly reflect the extent to which a city can solve problems before adjustment and renewal. It emphasizes the sufficient capacity of a city to ensure the local residents can maintain normal production and life when the disasters happen without outside rescue.

In the new century of information technology, the prevention and mitigation of urban civil engineering requires not only the study of structural disaster resistance performance, but also the use of BIM, GIS and other technologies to build an electronic information system for disaster prevention and mitigation of urban civil engineering. Only by combining these two aspects can we effectively improve the disaster prevention and mitigation capabilities of cities.

With the rapid development of informatization, cities in the developed countries are building urban disaster prevention and mitigation information systems based on the resilient city theory, and the role of urban disaster prevention and mitigation information systems in disasters is also fully demonstrated. For example, Japan is a country with frequent natural disasters. However, Japanese have learned from the frequent earthquakes in history, at present, they have built a more advanced urban disaster prevention and mitigation system so that the loss of every natural disaster (especially the earthquake disasters), is gradually reduced to a minimum. On May 21, 2004, an earthquake measuring 6.7 on the Richter scale occurred in Algeria. It was only a few days apart on May 26, a magnitude 7 earthquake occurred in Japan. The earthquake in Japan had a larger magnitude but no deaths in Japan and less property damage, while Algeria the loss was heavy and more than 300 people died. This is due to the national measures for disaster prevention and mitigation in Japan. Japan has developed the "Early Assessment System for Earthquake

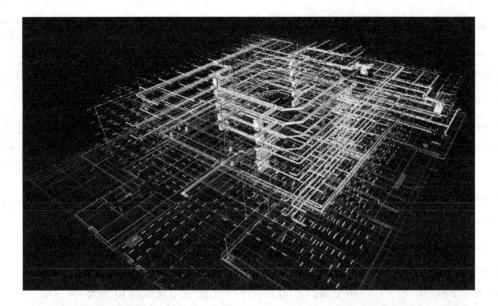

FIG.7-29 BIM and GIS

Disasters", which stores disaster data for earthquakes of magnitude 4 and higher. In the event of a major earthquake, the system can predict the scale of the disaster, so that the government can quickly take rescue measures. This fully demonstrates the importance of urban disaster prevention and mitigation information systems to the city against natural disasters.

Questions

- What are the types of disasters?
- Briefly describe the damage form of masonry structures, reinforced concrete structures

and large span space structures after earthquakes.

- The Beaufort wind scale is divided on what criteria?

- What are the reinforcement methods for reinforced concrete structures, masonry structures and steel structures?

- Describe the new achievements and development trend of disaster prevention and reduction.

Reference List

[1]   Ye Zhiming. Introduction to Civil Engineering[M]. Beijing: Higher Education Press, 2016.

[2]   Cui Jinghao. The severity of disasters and the importance of civil engineering in disaster prevention and mitigation [J]. Engineering Mechanics, 2006, 23:49-77.

[3]   Wang Ru. Civil Engineering Disaster Prevention and Mitigation[M]. Beijing: China Building Materials Industry Press, 2008.

[4]   Xu Zhenkai, Yuan Zhijun, Hu Jiqun. The method of building structure detection and reinforcement[J]. Engineering Mechanics, 2006, 23:117-130.

[5]   C S Holling. Resilience and Stability of Ecological Systems[J]. Annual review of ecology and systematics, 1973, 4: 1–23.

[6]   Liu Hengjun, Wang Kun, Huang Min. The discussion of digital information system for disaster prevention and reduction in civil engineering[J]. China Construction, 2009, 8: 80–81.

CHAPTER 8
APPLICATION OF INFORMATION
TECHNOLOGY IN CIVIL ENGINEERING

With the competition increasing in the construction markets and globalization process of the construction market accelerating, information technology (IT) was used in civil engineering as a support. Sharing civil engineering information resource, breaking the limitations of regional civil Engineering development, improving the management capacity and management level, and enhancing the competitiveness of the civil engineering enterprises, but also for all level of government and enterprises to establish a clean, transparent, and efficient management mechanism platform, thus achieving civil engineering technology quickly. This is bound to have a profound impact on the national economy and social progress.

8.1 Computer-aided Calculation

Definition: Computer-aided calculation is the use of computers and their graphics devices to help designers perform calculations and design work. In engineering and product design, computers can help designers with tasks such as calculation, information storage and drawing, which can reduce the labor of designers and shorten the work design cycle and improve design's quality.

8.1.1 Definition of computer-aided calculation

Computer-aided calculation is a method and technology that uses computer hardware and software systems to assist people in designing and calculating products and engineering.

Using computers, people can carry out **Computer Aided Design** (CAD), **Computer Aided Make** (CAM), **Computer Aided Engineering** (CAE), **Computer Aided Processing Planning** (CAPP), **Product Data Management** (PDM), **Enterprise Resource Planning** (ERP) and so on.

8.1.2 Significance of computer-aided calculation

- **Information management**

Computer provides the information support for the civil engineering construction, which ensures the construction of civil engineering effectively to some extent. The computer program mainly uses the computer information management software that carries on the systematic information management to the civil engineering construction equipment, the engineering design, the engineering system, the engineering quality, the engineering budget, the project cost budget, the engineering contract and the construction personnel and so on.

- **Accurate structure calculation**

The computer can adapt to the current complex structural calculation requirements, the data preparation workload is small, and various factors can be considered in the calculation, and the construction drawing is convenient.

When manually modifying the reinforcement, it should be able to directly select the component on the plane and modify it with intuitive sketches (such as PKPM, SASCAD and GSCAD), and directly modify the formed chart (such as GSCAD), and various plots. The data linkage between the modes makes it unnecessary for the structural engineer to perform supplementary calculations after the overall calculation, which reduces the workload.

- **Construction technology control**

In civil engineering construction, using computer to control construction technology is one of the effective ways to improve construction quality.

In the construction process, use the computer to carry on the automation control to the construction equipment, carry on the effective control management to each construction technology, and strive to realize the automation of the construction process.The overall measurement data and operation data of engineering equipment are analyzed with computer files, and the conclusions are provided with data support for construction.

Using the computer to control these time technology, greatly reduce the construction cost, on the basis of ensuring the overall quality of the project, accelerate the progress of the project, optimize the construction plan, and improve the economic benefits of the construction enterprise.

8.1.3 Application of computer-aided calculation

This section will focus on the functions and applications of several computer aided computing software.

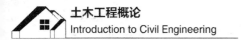

- **PKPM system**

The application of computer-aided computing in China in the field of civil engineering is the PKPM system developed by the China Academy of Building Research.

Types: The system is divided into reinforced concrete frame, frame frame, continuous beam structure calculation and construction drawing software (PK) and structural plane computer aided design software (PMCAD).

➤ **Main features:**

PKPM is a series. In addition to the integrated CAD system designed for construction, structure, equipment (water supply and drainage, heating, ventilation, air conditioning, electrical), PKPM also has a budget estimate for the building (reinforcement calculation, engineering calculation, engineering pricing), construction series software (bid series, safety calculation series, construction technology series), construction enterprise informationization (currently many nationally qualified enterprises are using PKPM information system).

➤ **Two major functions:**

- **Reinforced concrete frame, frame frame, continuous beam structure calculation and construction drawing software (PK)**

The PK module has two functions: two-dimensional structure calculation and reinforced concrete beam and column construction drawing.

The module itself provides a structural calculation software for the planar bar system. It is suitable for various rules and complex types of frame structures, frame-and-frame structures, and frame structures in industrial and civil buildings. The wall frame structure and the continuous reduction of the shear wall are continuous. Beams, arched structures, trusses, etc.

In the entire PKPM system, PK undertakes the work of assisting design of reinforced concrete beams and column construction drawings. In addition to the two-dimensional calculation results of the relay PK, the construction drawing assisted design of the reinforced concrete frame, the frame and the continuous beam can be completed, and the multi-level three-dimensional analysis software TAT, SATWE, PMSAP calculation results and the brick-concrete bottom frame and frame beam can be completed.

Perform strong column weak beam, strong shear weak bend, node core, column axial compression ratio, calculation and verification of column volume stirrup ratio, and also calculate elastic-plastic displacement and vertical seismic force calculation of weak layer under rare earthquake, frame beam crack width calculation, beam deflection calculation.

- **Structural Plane Computer Aided Design Software (PMCAD)**

PMCAD is the core of the whole structure CAD. The whole building structure model is the pre-processing part of PKPM two-dimensional and three-dimensional structure calculation software, and it is also the construction drawing design software and basic CAD of beams, columns, shear walls and floors.

PMCAD can automatically carry the load conduction from the slab to the secondary beam, the secondary beam to the load-bearing beam and automatically calculate the structural weight. It automatically calculates the load input by the human-computer interaction mode to form the load database of the whole building. The user can query and modify any part at any time.

- **YJK Software**

YJK Software is a building structure design software for the international market, including **YJK Building Structure Calculation Software** (YJK-A); **YJK Foundation Design Software** (YJK-F); **YJK Masonry Structure Design Software** (YJK-M); **YJK structural construction drawing aid design software** (YJK-D).

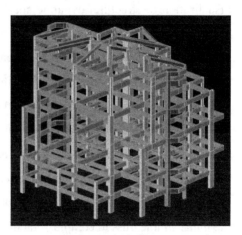

YJK design software system is a new set of integrated building structure aided design system, which includes six aspects: structural modeling, superstruc-

FIG.8-1　PKPM model diagram

ture calculation, foundation design, masonry structure design, construction drawing design and interface software.

The following is a model of a library project model established by YJK:

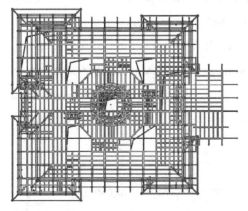

FIG.8-2　Plan of a library engineering model

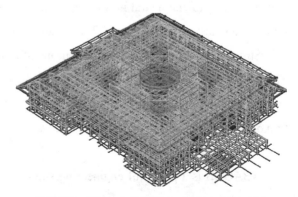

FIG.8-3　3D map of a library engineering model

- **MATLAB**

MATLAB is a high-level computer language developed by Math Works in the United States for scientific computing, data visualization, and interactive programming. Today MATLAB has extended its functions to many fields of scientific research and engineering applications.

a. Application of MATLAB in structural analysis

Definition: The structural analysis in civil engineering mainly refers to the solution of the internal force and displacement of the structure under the action of static and dynamic loads.

Due to the complexity of the structure and the strict requirements for the accuracy of the solution, the finite element method is a commonly used analytical tool. It is easier to realize the visualization of structural dynamic analysis in MATLAB, which is a new way and method of structural dynamic analysis.

b. Application of MATLAB in structural optimization

As an artificial intelligence algorithm, genetic algorithm is widely used in optimization analysis, but when using genetic algorithm, it needs to complete optimization calculation and structural analysis at the same time. It has the form of optimization function and the user's ability to set the algorithm and parameters. It is convenient to interface with FORTRAN or C.

Therefore, in the field of civil engineering, researchers have applied MATLAB optimization toolbox to solve specific optimization problems. The range of applications has also evolved from the initial simple truss to complex practical engineering.

c. Application of MATLAB in structural intelligent control and simulation

Intelligent control is inseparable from the establishment of artificial neural network, and MATLAB has its unique advantages in neural network. Based on MATLAB neural network technology in bolt support, tunnel engineering, structural foundation selection, material strength prediction and pile foundation bearing capacity Forecasts and other aspects have begun to apply. Although the artificial neural network system based on MATLAB has not been applied in structural damage testing and diagnosis, it is believed that it will be realized in the near future.

Structural simulation analysis is a means of replacing experimental research in some cases. The SIMULINK toolbox in MATLAB is a software package integrating modeling, simulation and analysis. It has unique advantages in dynamic system simulation and is suitable for simulation of structural vibration under earthquake action.

- **ABAQUS**

ABAQUS is a powerful engineering simulation software based on finite element method. It contains a very rich unit model, material model and analysis process. It has excellent ability to solve highly nonlinear problems and has strong applicability to the civil engineering industry. The use of ABAQUS software for CAE finite element analysis of building structures is currently a development direction of architectural design and plays an increasingly prominent role in architectural design and optimization.

a. Application in geotechnical engineering

ABAQUS has a constitutive model that can truly reflect the soil traits. It can calculate effective stress and pore pressure. It has powerful contact surface treatment function to simulate the phenomenon of disengagement and slip between soil and structure. Or the ability to excavate specific problems in geotechnical engineering, such as the initial stress state can be flexibly and accurately established.

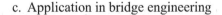

FIG.8-4   Tunnel Excavation Simulation of Underground Engineering of Beijing Guomao Subway Station

b. Application in building structural engineering

For the highly nonlinear material of concrete, ABAQUS provides four concrete constitutive models to adapt to different analysis environments, to achieve structural mechanics analysis, high-rise building structure formation analysis, high-rise shear wall elastoplastic dynamic analysis, seismic response analysis of frame structures, analysis of asphalt concrete cracking, and sound field analysis of building structures.

c. Application in bridge engineering

The bridge structure involves geometric non-

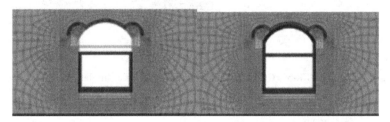

FIG.8-5   Erosion of oil tunnel walls

linear problems, which are caused by the interaction between large displacement, bending moment and axial force. ABAQUS has unparalleled advantages in dealing with nonlinear problems.

It has been used in reinforced concrete bridges, combined bridges. The construction process of prestressed concrete box bridges, large-span bridges, hydration heat analysis of mass concrete and underground structures are generally applicable.

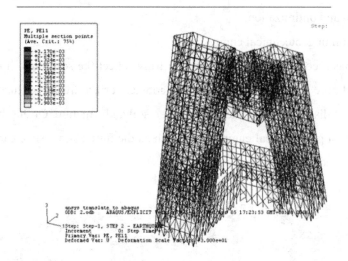

FIG.8-6　CCTV building nonlinear seismic simulation

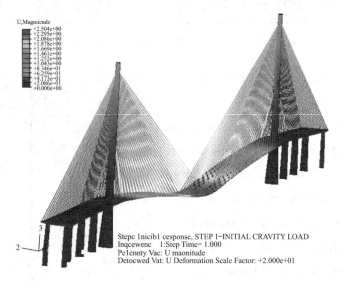

FIG.8-7　Analysis of the influence of gravity on the static stiffness of Qingpu bridge in Hong Kong

d. Application in water conservancy projects

ABAQUS can solve the problems of water pressure, temperature field, seepage field, gravity field and coupling of temperature field and force field, seepage field and force field in water conservancy engineering, etc.anti-seepage and underwater impact analysis, etc.

- **ANSYS**

a. Structural static analysis

Static analysis in ANSYS program not only performs linear analysis, but also nonlinear analysis such as plasticity, creep, expansion, large deformation, large strain, and contact analysis.

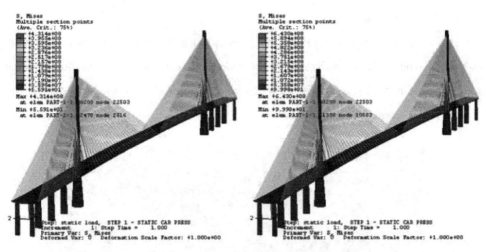

FIG.8-8　Analysis of the influence of vehicle on the static stiffness of Qingpu bridge in Hong Kong

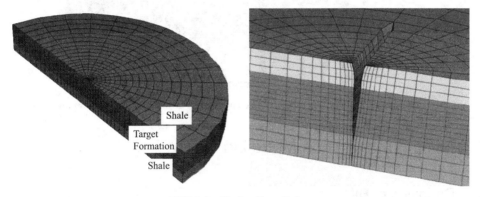

FIG.8-9　Hydraulic splitting

b.　Structural dynamics analysis

Structural dynamics analysis is used to solve the effects of loads over time on structures or components. The types of structural dynamics that ANSYS can perform include: transient dynamics analysis, modal analysis, harmonic response analysis, and random vibration response analysis.

c.　Structural nonlinear analysis

ANSYS program solves static and transient nonlinear problems, including material nonlinear, geometric nonlinear and element nonlinear.

d.　Dynamic analysis

ANSYS program can analyze large three dimensional flexible body motion. When the accumulation of motion plays a major role, these functions can be used to analyze the motion characteristics of complex structures in space and determine the stress, strain and deformation generated by the structure.

⑤ Fluid dynamics analysis

ANSYS fluid unit can perform fluid dynamics analysis, the type of analysis can be transient or steady state. The analysis results can be the pressure at each node and the flow rate through each cell. And the post processing function can be used to produce a graphical display of pressure, flow rate and temperature distribution.

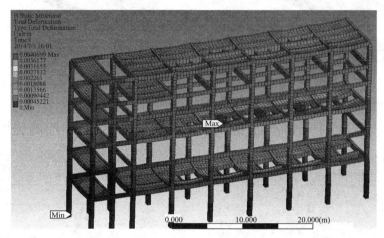

FIG.8-10   ANSYS structure analysis diagram of reinforced concrete structure

8.2   Building Information Modeling (BIM)

8.2.1   Definition of BIM

Charles Eastman, a professor at the Georgia Institute of Technology in the United States, proposed the basic content of BIM theory in 1975 under the name Chuck Eastman. In 2002, Autodesk published a white paper titled "Building Information Modeling," which creatively presented the term "Building Information Modeling."

Building Information Modeling is a complete information model that integrates engineering information, processes and resources of engineering projects at different stages of the life cycle into a single model, which is easily used by all project participants.

The definition of BIM by the US National BIM Standard (NBIMS) consists of three parts:

BIM is a digital representation of the physical and functional characteristics of a facility (construction project);

BIM is a shared knowledge resource that is a process of sharing information about the facility and providing a reliable basis for all decisions in the facility's life cycle from construction to demolition;

At different stages of the project, different stakeholders insert、 extract、 update and modify information in BIM to support and reflect collaborative work of their respective responsibilities.

8.2.2　Characteristics of BIM

- **Visualization**

The real use of visualization in the construction industry is very large. The visualization mentioned by BIM is a kind of visualization that can form interaction and feedback between the components.

In the BIM building information model, all the visual results can be used not only for the display of the renderings and the generation of the reports, more importantly, the communication, discussion and decision-making during the design, construction and operation of the project are all carried out under visual conditions.

- **Coordination**

The BIM building information model can coordinate the collision problems of various professions in the early stage of building construction and generate coordination data.

Certainly the coordination role of BIM is not only able to solve the collision problem between various professions. It can also solve, for example, the coordination of elevator shaft layout with other design layouts and clearance requirements, the coordination of fire zones and other design arrangements, the underground drainage arrangements, etc. arrangement and coordination of other design.

- **Simulation**

Imitation is not just a simulation of a designed building model. BIM can also simulate things that cannot be manipulated in the real world. For example, in the design stage, BIM can perform simulation experiments such as energy-saving simulation, emergency evacuation simulation, and sunshine simulation; in the bidding and construction phase, 4D simulation can be performed, that is the actual construction is simulated according to the organizational design of the construction. At the same time, 5D simulation can be carried out to realize cost control; in the later operation stage, simulations of daily emergency situations can be simulated, such as earthquake personnel escape simulation.

- **Optimization**

Optimization is constrained by three factors: information, complexity, and time. The BIM model provides information on the actual existence of the building, including geometric information, physical information, and rule information, as well as the actual existence of the building after the change. BIM optimization can do the following things:

① Project plan optimization: Combine project design and investment return analysis, and the impact of design changes on investment returns which can be calculated in real time.

② Design optimization of special projects: Optimize the special-shaped design such as podium, curtain wall, roof, large space, etc.and focus on optimizing the construction difficulty and construction problems, which can bring significant time and cost improvement.

- **Descriptive**

BIM can help the owner to produce the following drawings:

① Integrated pipeline diagram (after collision inspection and design modification, eliminating corresponding errors);

② Comprehensive structure hole map (pre-embedded casing diagram);

③ Collision check debugging report and suggestion improvement plan.

- **Integration**

Based on BIM technology, it can carry out the integrated management of the whole life cycle of the project from design to construction to operation. The core of BIM technology is a database formed by a three-dimensional computer model, which not only contains the design information of the building, but also can accommodate the whole process information from design to completion and even the end of the use cycle.

8.2.3　Application of BIM

CAD technology is a revolution in the field of design, and BIM technology will be a revolution in the field of civil engineering.

Through a highly integrated, digital and accurate BIM model, it can integrate design, construction companies, material suppliers, real estate developers, etc. into a digital management platform to achieve precise management and control, and achieve the full life of civil engineering, All-round computerized management. Building information models can be widely used in various stages of civil engineering, such as design, construction, operation and maintenance.

- **Parameterization**

Definition: Parametric modeling refers to the establishment and analysis of models by parameters rather than numbers. Simply changing the parameter values in the model can establish and analyze new models.

BIM models are presented in the form of components, between these components. The difference is reflected by the adjustment of the parameters, which preserves all the information of the primitive as a digital building component.

By inputting the settings including the height, size, material and other parameters of the component, the BIM-related software (such as Lumion, 3DMax) BIM greatly advances the progress of the informatization reform of the building engineering, and the engineering can be carried out through the three-dimensional building model. All the information of the project is displayed in detail.

FIG.8-11   Facade rendering of a teaching building

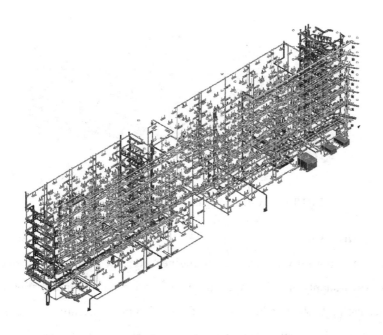

FIG.8-12   A model of the electromechanical network of a teaching building

● **Construction simulation**

BIM technology's visibility, simulation and synergy enable multiple professional cooperation and information exchange, saving engineering construction costs and speed.

Simulate the construction process, prepare a scientific construction organization plan, strictly observe the schedule of space-time construction, and rationally allocate the site and resources. It is convenient for the staff to comprehensively understand the progress of each stage of the project and promote the completion of the project schedule on time.

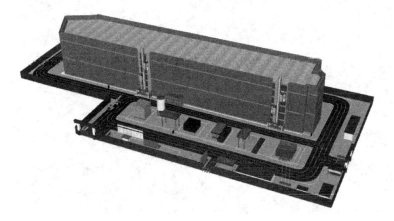

FIG.8-13 Construction site layout of the main stage of a teaching building

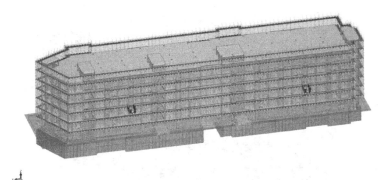

FIG.8-14 A scaffolding model of a teaching building

● **Collaborative work**

BIM technology not only involves civil engineering, but also involves electromechanical, municipal, water supply and drainage, etc. It summarizes all the information of the project through BIM technology, which is beneficial to the staff to make scientific decisions and is conducive to the cooperation of personnel in the construction process and to unified management.

Construction simulation inspections can be carried out before construction, reducing construction risks and saving on-site materials and equipment costs, and timely discovering problems in construction. The following figures show the results of collisions between equipment lines for water supply and drainage and electrical engineering in a project and modified by BIM technology.

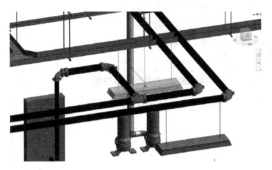

FIG.8-15   Revit check for water and electricity line collision problems

FIG.8-16   Revit changes the water supply line collision problem

- **Progress management**

Use BIM platform to carry out simulation construction, carry out construction conflict analysis through 5D model, and formulate corresponding management measures, combine construction progress management plan to on-site construction progress management, and make current construction progress and capital resource progress through building model clear The display is convenient for managers to understand the deviation between the current construction schedule and the plan, and to dynamically adjust the progress in time.

FIG.8-17   BIM5D construction process simulation

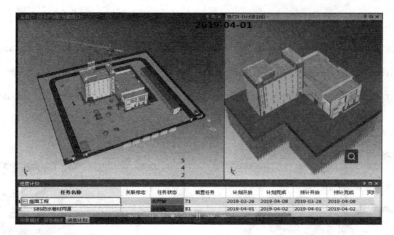

FIG.8-18 Dynamic management of the construction progress of an administrative building

8.3 Information Construction

8.3.1 Definition of information construction

Definition: Information construction refers to the extensive application of computer information technology in all stages of the construction process, collecting, storing, processing and communicating information on schedule, manpower, materials, machinery, funds, schedules, etc., and scientifically comprehensive utilization provides timely and accurate decision-making basis for construction management.

The construction unit shall follow the requirements of engineering design to formulate and implement the specific project construction and monitoring plan. Based on the monitoring results, timely adjust and scientifically optimize the construction plan and related processes, scientifically correct and rationalize the design according to information feedback and change.

8.3.2 Advantages of information construction

- **Improve the speed of information exchange**

In the process of management informationization in the construction industry, the use of information networks as a carrier of information exchange has greatly improved the speed of information exchange. The progress of the project can be inquired at any time to identify and solve existing problems in time.

Moreover, it can provide complete and accurate historical information, which provides great convenience for the parties involved in browsing information, and the efficiency of project management has been greatly improved.

- **Collect project information in time**

The management informationization of the construction industry can adapt to the requirements of large-scale use of project management information, and the information of management activities of each project should be collected every day, so that timely and comprehensive supervision of all management links in the construction project can be realized to improve the project. The quality of management work.

- **Improve project communication efficiency**

Engineering projects often involve multiple project participants. Due to the decentralization of project participants, all parties consume a lot of manpower, material resources and financial resources in communication. Through information sharing, information can be transmitted to many people at the same time, and often because of the limitation of time and space, communication at a long distance (such as international) is more convenient, and the interaction and speed of information communication are improved.

- **Realize project collaborative work**

In the environment of information sharing, by automatically completing some regular information notifications, the number of times that project participants need to communicate by humans can be reduced, and the transmission of information becomes fast, timely and smooth.

In addition, project information management is conducive to changing the traditional communication methods of the parties involved in the project, achieving orderly online collaborative operations, making communication and decision-making among project members consistent and synergistic, and better achieving the overall project aims.

- **Promote scientific decision-making and reduce engineering risks**

Establish a project information management system platform, use the information network as a carrier for project information exchange, speed up the exchange of project information, realize timely and accurate collection of project data, and systematically and structurally integrate all information from the construction site to the management level.

The management is able to check the progress of the project in time. Thereby providing quantitative analysis data for project management, supporting scientific decision-making of the project, and promoting effective and rapid prediction, analysis, prevention and control of various risks.

8.3.3 Deficiencies in information construction

- **The cost is difficult to grasp and control**

Because of the isolation between projects in different regions, it is difficult to formulate their own cost plans based on the financial cost of the law, resulting in the isolation of cost information.

In addition, the construction period of the general engineering project is long, and the long-term information closure makes it difficult for the general enterprise to grasp and control the construction cost of the project as a whole.

- **Isolation of construction management informatization**

With the increase of project scale and technical complexity, and the refinement of engineering projects, there are often dozens of construction companies involved in a large-scale project. In the construction information management, various units have various software compatibility. The problem hides the sharing and automatic exchange of data information resources, causing the phenomenon of "information islands", which seriously hinders the process of informationization.

The construction site management and the project management progress and quality control of the project in other places did not achieve good interactivity, which is not conducive to the overall project management of the construction enterprise.

- **Insufficient investment in information construction**

The investment in enterprise informatization is obviously insufficient, and the output value of enterprises accounted for only 0.027% on average, which is far from the 0.3% of developed countries. The profits of construction enterprises are not high and the proportion of hardware and software input structure is out of balance.

8.4 Future of Civil Engineering IT

8.4.1 Application directions

With the in-depth development of the Internet, civil engineering information technology will develop in the direction of network, intelligence and integration.

- **Development of databases and management systems**

The development of databases and management systems in civil engineering information technology is the main development trend of civil engineering information technology in the future. The database and the management system under the internet technology mainly through the conceptual definition of the relatively abstract data model and the generalization of the decentralized implementation, under the civil engineering information technology, the stored related information level is higher, and the query is made. It is more convenient and quick, and has the right to manage and operate the corresponding data effectively, so that the corresponding system under civil engineering information technology can recover more quickly after being subjected to abnormal conditions.

- **Application of new computer aided systems**

The application of new computer aided systems is an important development trend in the future of civil engineering information technology. Computer technology has been continuously researched and developed. Computer-aided technology analyzes and scientifically rationalizes the corresponding structural performance and data of more complex engineering and products in civil engineering through computer networks.

The development of new computer-aided systems has made significant progress at this stage. Under the influence of computer technology and network technology, the new computer-aided system can be effectively applied to the analysis of civil engineering materials and the rational construction of civil engineering structures. It plays an important role in the development of civil engineering information technology.

- **Development of 3D technology and virtual technology**

The application of 3D technology and virtual surgery is a key development trend in the future of civil engineering information technology. In the new era, three-dimensional technology and virtual technology are the products of civil engineering information technology in the process of scientific and technological development.

The development of 3D technology and virtual technology is mainly through the use of a new computer-aided system to set the 3D model of the corresponding product, and then through the civil engineering information interaction software to build the program to interact, so that users can pass the mouse and other equipment Take the appropriate action.

8.4.2　Application areas

Here introduces several new applications of information technology in civil engineering field:

- Bunker Sponge City

Application of Information Technology in Sponge City Construction

① Seepage, stagnation, storage, net, use and discharge

For the layout of "seepage, stagnation, storage, net, use and drainage" measures and the application of technology, first of all, to deal with the "seepage, stagnation, storage, net, use, drainage" between the relationship, storage, hysteresis, seepage to reasonably plan the proportion of air storage (roof garden), surface stagnation, soil infiltration, underground storage, study the coupling of rainfall characteristics and stagnation, storage, drainage, on the basis of the design of the best matching scheme with the drainage system, and scientific evaluation, so that it can meet the requirements of flood control and drainage, so as not to waste too much investment. Through the construction of water system connectivity project and node control project to restore and strengthen the hydraulic connection between the catchment area and the cavernous body.

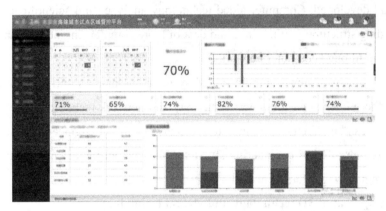

FIG.8-19 Sponge City Scheduling Management Platform
(www.sinfotek.com)

② Connecting technology of urban drainage system

Urban drainage and waterlogging prevention involves source control, rainwater pipe network (small drainage), waterlogging prevention (eg.urban open channel, depressions, inland waterways, regulation and storage, belonging to the upper reaches of large drainage) and drainage projects (eg.river channels, drainage pumping stations, etc. belonging to the lower reaches of large drainage).

● Intelligent construction

Several Applications of Information Technology in Intelligent Site Construction:

① Intelligent spray dust control

With the help of intelligent equipment to carry out intelligent spray dust control on construction sites, reduce the environmental pollution in the construction process to a minimum, and implement corporate social responsibility. The intelligent system has the advantages of energy saving and high efficiency and low cost, it can realize the uninterrupted work of 24 hours, and realize the real-time transmission of relevant data to the computer. Through the effective analysis of the index data, the staff take effective measures to control the relevant work of the site in time, improve the environment of the construction site and ensure the health and safety of the construction personnel.

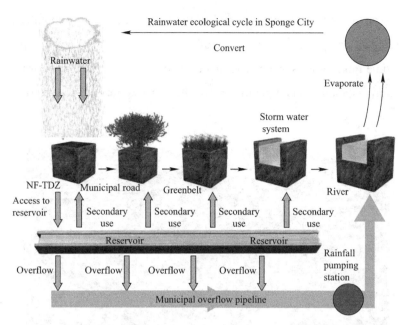

FIG.8-20 Sponge City Drainage Connection Technology

② Daily management software

Applying to the schedule management, we can supervise every link of the construction site at any time, summarize the progress of each project in real time, and divide the various links of the project in a planned way, including the responsible person, the manager, the time of completion of the project and the reasons for the delay, so as to facilitate the first

FIG.8-21 Intelligent Spray Dustfall

time to deal with and solve the problem in a timely manner.

③ Labor personnel management activities

With the help of the intelligent site system, the management of labor services can be realized by using the advantages of big data in information processing, from real-name registration, real-time attendance to intelligent location identification to improve the scientific nature and effectiveness of employment management. The intelligent site system can also realize the intelligent safety helmet positioning system and mobile phone application positioning, which can deal with some unexpected situations in time and reduce the cost of risk management.

④ Tower crane safety monitoring system

Tower crane safety is the object that needs to be paid more attention to in construction. The intelligent construction site system is applied to the tower crane safety monitoring system to realize the intelligent and practical monitoring of engineering management, equipment management and real-time monitoring system. The main function of the safety monitoring system is to monitor the safety data and related parameters of the construction site tower crane in real time and dynamically.

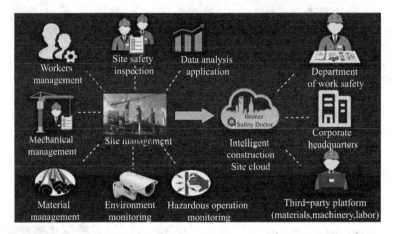

FIG.8-22 Smart Site Management

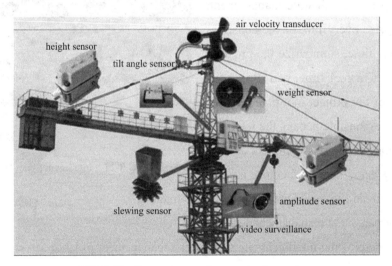

FIG.8-23 Tower crane safety monitoring system

Questions

- In addition to the five computer-aided design software introduced in the first section, can you list a few other examples?
- Briefly outline the definition and characteristics of BIM.
- The application of BIM in the whole life cycle of the project, in which part do you think

BIM is the most practical?

- Advantages and disadvantages of information construction.

- Which technologies are the focus of future applications of civil engineering information technology?

- Can you give another example of the application of information technology in civil engineering?

Reference List

[1] Chen Fang. Development and Construction of Information Technology in Civil Engineering[J]. Home Business, 2015(22): 156-158.

[2] Shi Yushun. Discussion on the information construction and prospect of civil engineering design[J]. Engineering Technology (full text), 2016(12):00089-00090.

[3] Zhang Yusheng. Difficulties and countermeasures of management informationization in construction industry [J]. Zhonghua Dwelling, 2011 (06): 129.

[4] Yin Hongliang. Application and Development of Informatization in Construction Engineering Management. [J]. Shenzhou Mid Issue, 2019(02): 267.

[5] Wang Wei. Difficulties and Countermeasures of Management Informationization in Construction Enterprises[J]. City Construction Theory Research: Electronic Edition, 2013(28).

[6] Ji Liangwei. Application Analysis of BIM Technology in Bridge Construction[J]. Enterprise Technology and Development, 2018(06): 148-149.

[7] Deng Dehai. Application and Discussion of BIM Technology [J]. China Real Estate Mid, 2017(16).

[8] http://space304535.ccbuild.com/portal.php?mod=view&aid=601473.

[9] Wang Jinchang. Application of ABAQUS in Civil Engineering[M]. Hangzhou: Zhejiang University Press, 2006.

[10] Chen Xixiang. Application of MATLAB technology in civil engineering field [J]. Electronic World, 2017 (5): 142.

[11] Bi Jianwen. On the Role of Computer in Civil Engineering Construction[J]. China Electronic Commerce: Science and Technology Innovation, 2014(14):11-11.

[12] Fang Zheng, Zhang Lei, Liu Fei. Analysis of sponge city construction and related technical problems [J]. Journal of Hubei Institute of Technology, 2016(3):31-36.

[13] Shi Jinhao, Liang Lei, Shi Tailong. Application of Intelligent Site System in Construction Process [J]. Southern Agricultural Machinery, 2019(11):203-203.